Molecular Diagnosis

S. Jeffery,[a] † J. Booth[a] and S. Myint[b]
[a]St George's Hospital Medical School, London, UK
[b]Department of Microbiology, University of Leicester, Leicester, UK

BIOS

© BIOS Scientific Publishers Limited, 1999

First published 1999

A CIP catalogue record for this book is available from the British Library.

ISBN 1-85996-190-8

BIOS Scientific Publishers Ltd
9 Newtec Place, Magdalen Road, Oxford OX4 1RE, UK
Tel. +44 (0)1865 726286. Fax +44 (0)1865 246823
World Wide Web home page: http://www.bios.co.uk/

Published in the United States of America, its dependent territories and Canada by Springer-Verlag New York Inc., 175 Fifth Avenue, New York, NY 10010-7858, in association with BIOS Scientific Publishers Ltd.

Published in Hong Kong, Taiwan, Singapore, Thailand, Cambodia, Korea, The Philippines, Indonesia, The People's Republic of China, Brunei, Laos, Malaysia, Macau and Vietnam by Springer-Verlag Singapore Pte. Ltd, 1 Tannery Road, Singapore 347719, in association with BIOS Scientific Publishers Ltd.

This book is dedicated to my colleague Jim Booth and to my beloved nephew Oliver who both died before it was completed.

S. Jeffery

Production Editor: Andrea Bosher.
Typeset by Marksbury Multimedia Ltd, Midsomer Norton, Bath, UK.
Printed by Biddles Ltd, Guildford, UK.

Molecular Diagnosis

THE MEDICAL PERSPECTIVES SERIES

Advisors:

B. David Hames *School of Biochemistry and Molecular Biology, University of Leeds, UK.*

David R. Harper *Department of Virology, Medical College of St Bartholomew's Hospital, London, UK.*

Andrew P. Read *Department of Medical Genetics, University of Manchester, Manchester, UK.*

Oncogenes and Tumor Suppressor Genes
Cytokines
The Human Genome
Autoimmunity
Genetic Engineering
Asthma
HIV and AIDS
Human Vaccines and Vaccination
Antibody Therapy
Antimicrobial Drug Action
Molecular Biology of Cancer
Antiviral Therapy
Molecular Virology *Second Edition*
DNA Fingerprinting *Second Edition*
Understanding Gene Therapy
Molecular Diagnosis

Forthcoming titles:
Genetic Engineering *Second Edition*

Contents

[a]Contributed by Dr Howard Pringle, Department of Pathology, Leicester Royal Infirmary, Leicester, UK

Abbreviations

ALL	acute lymphoid leukemia
AML	acute myeloid leukemia
AP-PCR	arbitrarily-primed PCR
ARMS	amplification refractory mutation system
AS	Angelman's syndrome
CFTP	cystic fibrosis transmembrane protein
CFTR	cystic fibrosis transmembrane regulator
CGH	comparative genome hybridization
CMV	cytomegalovirus
CVS	chorionic villus sample
DAF	DNA amplification fingerprinting
ddATP	dideoxy adenosine triphosphate
ddCTP	dideoxy cytosine triphosphate
ddGTP	dideoxy guanosine triphosphate
ddTTP	dideoxy thymidine triphosphate
DEPC	diethyl pyrocarbonate
DGGE	denaturing gradient gel electrophoresis
DMD	Duchenne muscular dystrophy
DNA	deoxyribonucleic acid
DNA-Mtase	DNA methyltransferase
dNTP	deoxynucleotide triphosphate
ds	double stranded
EBV	Epstein Barr virus
ER	estrogen receptor
EtBr	ethidium bromide
FAP	familial adenomatous polyposis coli
FIGE	field inversion gel electrophoresis
FISH	fluorescent in situ hybridization
FMTC	familial medullary thyroid carcinoma
GI	guanidium isothiocyanate
HA	heteroduplex analysis
HBV	hepatitis B virus
HIV	human immunodeficiency syndrome
HLA	human leukocyte antigens
HNPCC	hereditary non polyposis colorectal cancer
HPA	hybridization protection assay
HPV	human papilloma virus
ISH	in situ hybridization

IVF	in vitro fertilization
LCR	ligase chain reaction
LOH	loss of heterozygosity
MEN	multiple endocrine neoplasia
MRD	minimal residual disease
mRNA	messenger RNA
MRSA	methicillin-resistant *S. aureus*
MVR-PCR	mini-satellite variant repeat-PCR
NASBA	nucleic acid sequence based amplification
NSA	non specific amplification
PCR	polymerase chain reaction
PCR-RE	PCR-restriction endonuclease
PCR-RFLP	PCR-restriction fragment length polymorphism
PCR-SSCP	PCR-single strand conformation polymorphism
PFGE	pulsed field gel electrophoresis
PTT	protein truncation test
PWS	Prader Willi syndrome
RDA	representational differential analysis
rDNA	ribosomal DNA
RFLP	restriction fragment length polymorphism
RNA	ribonucleic acid
rPCR	random PCR
rRNA	ribosomal RNA
RT	reverse transcriptase
RT-PCR	reverse transcriptase PCR
SDA	strand displacement amplification
SISPA	sequence-independent single primer amplification
SNRPN	small nuclear riboprotein
ss	single stranded
SSCP	single stranded conformational polymorphism
TAS	transcription based amplification system
TMA	transcription-mediated amplification
TRAP	telomeric repeat amplification protocol
tRNA	transfer RNA
YACs	yeast artificial chromosomes

Dedication

James Clive Booth, Jim to all his many friends and colleagues, died suddenly at home on 24th January 1998 at the age of 56 years. He was a Reader in Virology at St. George's Medical School.

Jim graduated with a degree in bacteriology from Birmingham University and worked initially in veterinary pathology at Cambridge University, where he isolated, in tissue culture, the virus responsible for inclusion-body rhinitis in pigs. From there he made his first association with St. George's by working for his PhD under Professor Stern, on the development of an inactivated vaccine against rubella. A year at the Karolinska Institute in Stockholm studying viral antigenicity, was followed by a period as a senior scientist at the Wellcome Laboratories where he used this knowledge to produce vaccines against foot and mouth disease. In 1975 he returned to St. George's to stay. He gained Membership, and then Fellowship, of the Royal College of Pathologists.

During his time at St. George's Jim was heavily involved in setting up and expanding the virus diagnostic laboratories, and was given honorary status as a clinical scientist. His primary research interest was in the role of cytomegalovirus, and this led to a study of the complex relationship of this virus with graft rejection. The involvement of coxsackie virus in heart disease was an area he became involved in shortly before his death.

His ability to incorporate academic research into clinically relevant areas, and an awareness of the importance of uniting the two disciplines, made him a valued member of the St. George's staff. He served on many committees and was always a popular chairman, as he had that precious gift of keeping discussion limited to what was necessary. Jim was forever available to give help and advice to both students and colleagues, and his good humor and approachability will be missed by all.

Away from St. George's, Jim was a very special man to his wife Pat and two sons, Martin and Ralph. He was an inspirational father and husband who blended good humor, common sense, and approachability to form a rich and valued personality. He was a great admirer of opera, fine literature, and experimental cookery! Despite having his own personal tastes, he took great interest in trying almost anything, from watching the Rolling Stones to hill walking in France. His enthusiasm, companionship, and sense of humor are sadly missed.

Preface

The object of this book is to show how the analysis of nucleic acids is used in the study of genetic and infectious disease. The major methodologies have been described, and then their uses in the different areas illustrated in chapters on genetics, bacteriology, virology, and histopathology. We have stepped very briefly outside these areas to show that the same techniques can be applied in the study of fossil remains. There are many other fields in which the study of nucleic acids is important, forensic science for example, but the scope of this book does not extend to discussing all of these. Although there are many variations on the themes of DNA and RNA analysis, the basic methods often remain the same, and we hope that this book gives sufficient information to understand how in other disciplines, similar procedures can be applied.

Acknowledgements

There are too many people who have helped indirectly or directly with this book to mention them all, so I hope the majority are not offended by the mention of a few. I would like to thank all my friends and colleagues in the Medical Genetics Unit at St. George's Medical School, for making it a place where it is easy to learn what goes on in both research and diagnostic laboratories. My special thanks go to Professor Nick Carter, without whose friendship and example I would never have been in, or stayed in, the field of Medical Genetics (this may or may not be a good thing!). Although only a short book there were many times when I thought it would never be finished because of the real tragedies that occurred along the way. Thank you to Christiane Fenske for suggesting suitable cover images. For ensuring that there was an eventual end product, a big 'thank you' to Rachel Offord, who was a Commissioning Editor of boundless patience and enthusiasm. Whether she was right to be so, only the reader can judge.

Chapter 1

Introduction

The aim of this book is to show how nucleic acids, that is deoxyribonucleic acid (DNA) and ribonucleic acid (RNA), can be used diagnostically. DNA is most commonly used for diagnostic purposes, as it is a very stable molecule. The variations between the DNA from different individuals means that it is possible to identify a specific person from their DNA. In the same way, different strains of bacteria can be identified, and mutations causing genetic disease can be distinguished from normal genes. Diagnosis using nucleic acids is not limited to medicine, but can be applied in such diverse fields as microbiology, genetics, oncology, forensic science, and even archaeology. Although the areas of use are different, the methodology is often similar. For that reason, the many methods that are commonly used in such analysis will first be described, followed by chapters on how they can be applied in different scientific disciplines. Methods that are primarily used in one particular discipline will be detailed in their respective chapters. Although an extensive range of methods are discussed, there are many more variations possible on each theme, and for these more detailed reviews of each method should be consulted.

Before describing the main methodologies in Chapter 2, a brief overview of nucleic acid structure and its organisation within mammalian cells and micro-organisms will be presented. DNA is composed of four bases, adenine (A), thymine (T), guanine (G), and cytosine (C). The bases are the same in RNA except that thymine is replaced by uracil (*see Figure 1.1*). The bases are attached to a backbone of sugar molecules, deoxyribose for DNA and ribose for RNA. DNA has a double stranded stucture, in which the bases bind very specifically, A to T and G to C (*Figure 1.2*).

1.1 Nucleic acids in human cells

There are about 3000 million bases in the human genome, of which 5–10% code for genes. The majority of the remaining DNA has no known function, and is often referred to as 'junk' DNA on the assumption that it is indeed functionless. However, the history of science is filled with many such assumptions which have proved unfounded, so it is probably better to leave a question mark over this 90% of the genome. There are

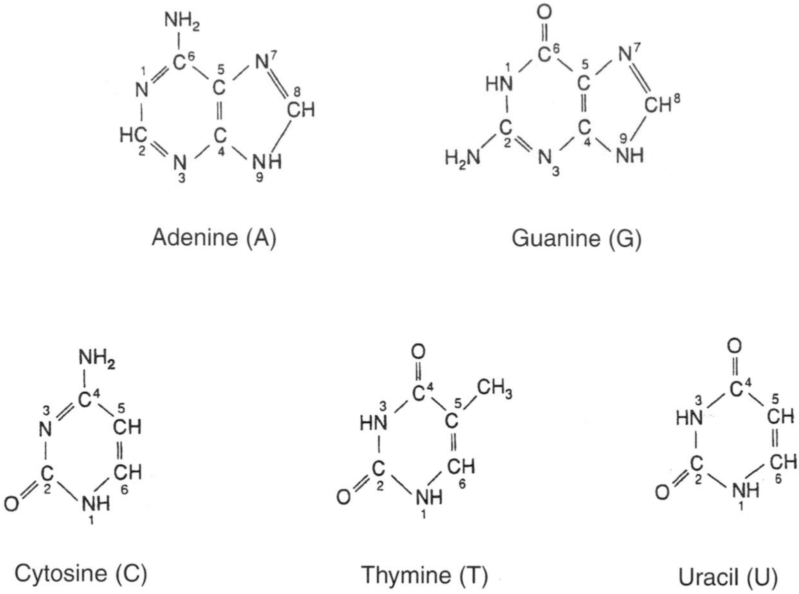

Adenine (A) Guanine (G)

Cytosine (C) Thymine (T) Uracil (U)

Figure 1.1: The bases that make up nucleic acid acids. Thymine in DNA is replaced by Uracil in RNA. (Reproduced from T. Strachan and A.P. Reed (1996) *Human Molecular Genetics* 1st Edn. BIOS Scientific Publishers, Oxford.)

estimated to be in the region of 75 000 genes in human DNA. The vast majority of human DNA is found in the cell nucleus, and this is where most of the information on human genetic disease in this book will be concentrated. However, there is also an important source of DNA in the mitochondria, which are found in the cytoplasm. These organelles have a different arrangement of genes compared to the nucleus (discussed below), and the mitochondrial genome contains only 37 genes. Owing to a high mutation rate in mitochondrial DNA and the fact that the mitochondrial genome is densely packed with coding sequences, this DNA is more important than might be expected in human genetic disease.

Human nuclear DNA is complexed with proteins in the form of chromosomes. During most of the cell cycle these structures are dispersed in the nucleus, but during cell division they condense and can be seen under the microscope with the correct preparative techniques. Normal human cells contain 22 pairs of chromosomes (called autosomes), one of each pair inherited from each parent, plus 2 sex chromosomes which are known as X and Y. Females are XX and males XY. *Figure 1.3* shows a diagram of the human chromosome complement, which are numbered in size order from 1 to 22. The chromosomes represented in *Figure 1.3* would have been treated with a stain called Giemsa, to give the light and dark pattern known as G banding. Light bands are much more gene rich than the dark bands. *Figure 1.3* also shows that chromosomes have 2 arms, with a constriction where they meet called the centromere. The short arm is the p arm (from the French petite), the long one is the q arm (because q

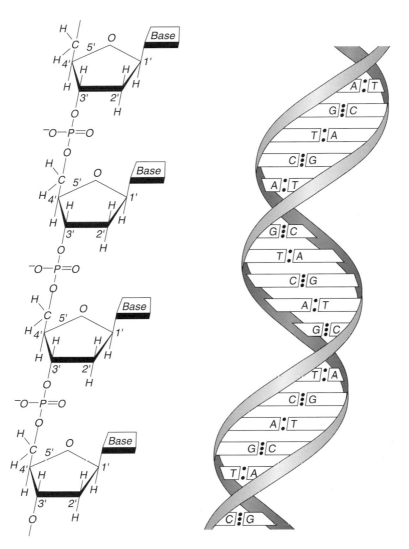

Figure 1.2: The bases as they occur in DNA with the deoxyribose sugar backbone (left), and the molecular structure of DNA (right). (Reproduced from (D.J. Weatherall (Ed.)) *The New Genetics and Clinical Practice*. 3rd edn. Oxford University Press, 1991.)

comes after p). The centromere is essential for segregation of duplicated chromosomes when they have divided at mitosis. The end of each chromosome arm is called a telomere, and this contains large numbers of repeated lengths of noncoding DNA. These are very variable in number between different individuals, and this variation is one component that is used to produce a 'DNA fingerprint'.

Human genes were once thought to exist as uninterrupted coding regions in the nucleus, which reflected exactly the messenger RNA (mRNA) sequence found in the cytoplasm. It is now clear that in all but a

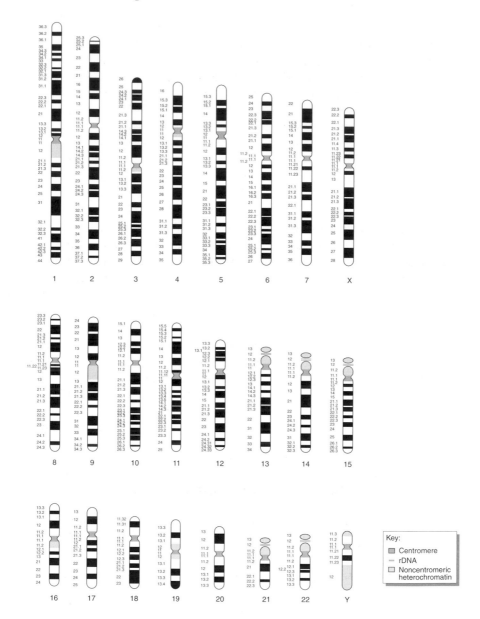

Figure 1.3: Banding pattern of human chromosomes (G-banding, 550-band karyogram). Note that chromosome 21 is in fact smaller than chromosome 22 in size. The observed metaphase chromosome lengths range between 2 μm (chromosome 21) and 10 μm (chromosome 1), whereas the fully uncoiled DNA strands would be expected to measure between 1.7 and 8.5 cm respectively. Note the presence of extensive heterochromatic regions on the Y chromosome, at the secondary constrictions of chromosomes 1q, 9q and 16q, and on the short arms of the acrocentric chromosomes 13, 14, 15, 21, and 22. (Reproduced from T. Strachan and A. P. Reed (1996) *Human Molecular Genetics* 1st Edn. BIOS Scientific Publishers, Oxford.)

very few cases this is not true. Most genes are interrupted by noncoding regions as summarized in *Figure 1.4*. RNA is copied from DNA, and initially this RNA contains both coding and noncoding regions (*Figure 1.4*). The stretches of RNA which code for amino acids are called exons, and the intervening sequences introns. There are 2 bases at the beginning and end of every intron that are invariant; these are GT at one end and AT at the other. This intron/exon boundary is known as a splice junction, because during processing to produce the final mRNA the intron sequences are spliced out at these points (*Figure 1.4*). The production of a messenger RNA molecule from DNA is known as transcription (*Figure 1.4*), and in part this process is controlled by particular sequences of bases upstream of the coding region. These are indicated as 'promoter' regions in *Figure 1.4*. There is an 'initiation' codon, AUG, in mRNA, where translation into protein begins in the cytoplasm. Upstream of this site, a specially modified nucleotide, 7-methyl guanosine (termed a 'cap') is added immediately after transcription, and most mRNAs have a series of adenosine molecules added at their 3' end, the so-called 'poly (A) tail'. These are the processes referred to as capping and polyadenylation in *Figure 1.4*.

RNA is normally single-stranded, and there are three types of RNA molecule in a cukaryotic cell. The most important in terms of genetic disease is mRNA, which carries the coding sequence that the cell recognizes to translate into protein. This translation involves the other two RNA species; ribosomal RNA (rRNA) and transfer RNA (tRNA). The former is a component of ribosomes, where the mRNA is translated into protein, while the tRNA carries the amino acids to the ribosomes,

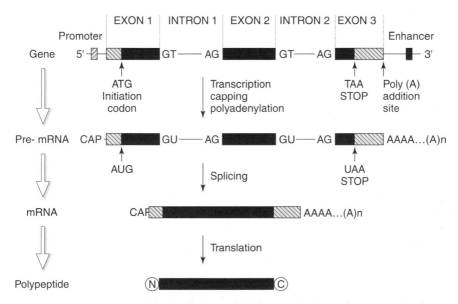

Figure 1.4: Structure and processing of a typical human gene.

and ensures that the correct amino acid is added as each triplet base code is read by the ribosomal complex.

1.2 Nucleic acid structure in micro organisms

1.2.1 Viruses

Viruses are genetic elements enclosed in a protective coat that allows them to move from one cell to another. The type of nucleic acid in a virus and the method of replication vary, and there are also varying complexities of DNA double helices in different viral species. Viruses may be subdivided by genome type, into those with:

- Double stranded (ds) DNA genomes (including the Adenoviridae), which are amongst the largest of viral genomes;
- Single stranded (ss) DNA genomes, typically small, such as parvovirus.
- dsRNA RNA genomes, (e.g., reovirus);
- ssRNA genomes, which can be subdivided into those that function as mRNA (positive sense), and those that are complementary to the mRNA produced from them (negative sense).
- Viruses with RNA genomes that use a DNA intermediate (a provirus) to produce the RNA genome (Retroviridae);
- Viruses with DNA genomes that use an RNA intermediate stage to produce a DNA genome (Hepadnaviridae).

The methods of replication of these various viruses are summarized in *Figure 1.5*.

The replication of RNA viral genomes occurs through the formation of complementary strands, a process that is catalyzed by RNA polymerase enzymes known as replicases. As mentioned above, there are so called negative strand RNA viruses (e.g., influenza) and positive strand viruses (such as polio). The difference is that in the former case the infecting viral RNA does not code for protein; it is only the complementary strand, and there must be a preformed replicase present to produce infectious strands. In the case of positive strand viruses the viral strand is coding and can act as a messenger RNA in the cell. Some viruses can replicate in the cytoplasm, or can incorporate into the host cell DNA. The RNA viruses that do this are called retroviruses, such as HIV, where part of the genome codes for an enzyme not present in higher cells; reverse transcriptase. This produces DNA from RNA, and in this way the viral RNA genome can be integrated into that of the cell. For more detail on viral genomes and their replication see *Molecular Virology* by D. R. Harper.

There are many forms of viral genome, with differing complexity of life cycles and replication. What most concerns the diagnostic laboratory is the nucleic acid that comprises the viral genome, since this will affect the techniques used. In those viruses where both infectious and noninfectious particles occur it is possible to distinguish between the two states.

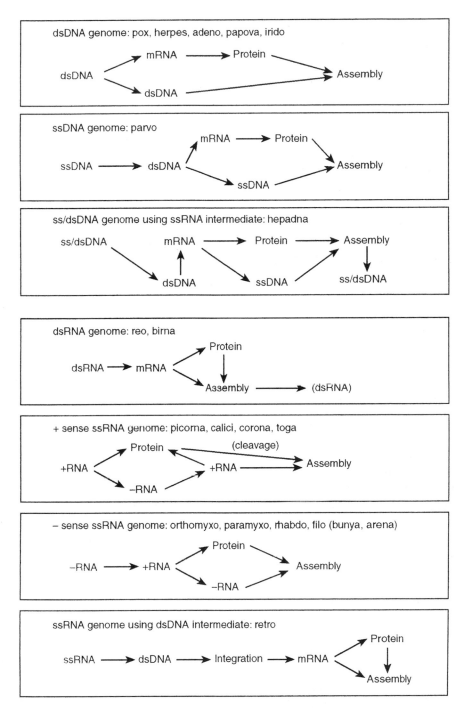

Figure 1.5: General methods of viral replication. (Reproduced from D.R. Harper (1998) *Molecular Virology*. BIOS Scientific Publishers, Oxford.)

1.2.2 Bacteria

Like higher organisms, bacterial genomes are composed of DNA, from which messenger RNA molecules are produced. There are no mitochondria and hence no mitochondrial DNA. However, there are DNA molecules within bacterial cells which can self-replicate. They are called plasmids, and have a small circular chromosome. Plasmid DNA does not usually encode genes with essential functions for the bacteria, which can therefore multiply quite happily without them, but they do carry resistance genes for antibiotics. There are often multiple copies in each bacterium, and their division occurs independently of the bacterial chromosome.

Different bacteria have different sized chromosomes, but *E. coli*, one of the best studied bacteria, provides a good example. It has a circular chromosome of 4 700 000 bases. This sounds a considerable number, but just one human gene, that for dystrophin (see Chapter 3) is almost half this size. What does illustrate the difference between bacterial and human genomes is the fact that there are three billion bases in human nuclear DNA, about a thousand times more than in bacteria. Unlike human cells, bacterial DNA is not enclosed in a nucleus, but it is centrally located within the organism. Another disimilarity is that the *E. coli* chromosome is packed with genes. About 90% of the bacterial DNA codes for messenger RNAs, compared to about 10% in humans. In bacteria, many genes with related functions, such as members of particular metabolic pathways, are clustered together in what are called 'operons'. These clusters are are transcribed as single messenger RNA molecules. There are no introns in bacterial genes, so the process of transcription from DNA to RNA is much simpler, with no splicing necessary to remove noncoding regions. Almost all *E. coli* genes occur as single copies, with the exception of those that code for ribosomal RNA, which in most strains of *E. coli* have seven copies. Genetic elements called transposons exist in bacterial DNA (as they do in eukaryotic DNA). These are DNA sequences that code for enzymes which can cause a new copy of the transposon to be inserted into another site of the chromosome. They can be simple or complex transposons; in the former case they contain only genes required for transposition, in the latter they can contain one or more bacterial genes between two simple transposon sequences.

Different bacterial species have different genomes, and these alterations can be exploited for nucleic acid diagnosis.

Further reading

Alberts, B., Bray D., Lewis, J., Raff, M., Roberts K. and Watson, J.D. (1994) *Molecular Biology of the Cell*, 3rd Edn. Garland Publishing, New York.

Harper, D.R. (1994) *Molecular Virology*. BIOS Scientific Publishers, Oxford.

Watson, J.D. and Crick, F.H.C. (1953) A structure for deoxyribose nucleic acid. *Nature* **171**: 737–738.

Chapter 2

Methods

The previous chapter gave a brief introduction to the types of nucleic acids in human cells and micro-organisms. In the following section some of the methods commonly used in nucleic acid diagnosis will be outlined, so that they can be referred back to in the chapters that follow. Methods that tend to be used mainly in one particular diagnostic area will be detailed in the appropriate chapters.

2.1 Hybridization detection methods

Before the invention of the polymerase chain reaction (PCR), described later, most nucleic acid sequences were detected using hybridization based on the specific nature of base pairing. The four bases in DNA bind very specifically, A to T and G to C (Chapter 1, *Figure 1.2*). It is this specificity which is used in hybridization methods of detection. Hybridization is the process where a piece of nucleic acid is incubated with its complementary strand, either in solution or with one strand on a solid support, and the buffer conditions are such that the two strands bind together by hydrogen bonding. Any nonspecifically bound DNA can be removed at subsequent washing steps, again using the appropriate conditions for buffer and temperature. Hybridization can be between DNA and DNA, DNA and RNA, RNA and RNA, or either nucleic acid and synthetic lengths of nucleotides, called oligonucleotides. Usually, the nucleic acid under investigation is bound to a support, such as a nylon membrane, while the specific probe is in solution. One of the most frequently used hybridization methods in diagnosis before PCR was Southern blotting (named after its inventor, Ed Southern) and this is described in Section 2.1.3.

Hybridization detection methods rely on the specificity of base pairing, as mentioned, and also on the fact that mismatched sequences that bind have a different melting temperature compared to those which are correctly matched. When heated, strands with sequences containing mismatches will separate at lower temperatures than those which match exactly. A mutation caused by a single base change can be detected using synthetic oligonucleotides complementary to the normal and mutated part of the gene, simply by using the correct washing conditions. To be visualized, the probes have to 'labeled' in some way (see Section 2.2).

2.1.1 Restriction fragment length polymorphisms

Human genomic DNA is far too large to examine in its entirety so it has to be cut into smaller fragments, which can then be separated according to their size. A common method of doing this is to use restriction enzymes. These are bacterial enzymes, believed to be part of the defence against viral infection. The enzymes recognize a specific sequence of bases in DNA, and cut the molecule where this sequence occurs. The enzymes are named from the bacteria from which they are isolated, for example, the enzyme EcoR1 came from *E. coli* strain R1, and it cuts wherever the sequence GAATTC is found in DNA. Other enzymes cut at different sequences, 4 to 8 (or more) bases long. A shorter recognition sequence produces more cuts in the DNA. The sizes of some fragments produced by these enzymes varies between different individuals, and these different sized fragments are known as RFLPs (restriction fragment length polymorphisms). This is discussed further, below. The detection of different sized fragments can be very important in analyzing certain genetic disorders, as will be explained in Chapter 3. Differences between individuals occur because about 90% of human DNA is noncoding, that is, does not contain genes, and variations in base sequence between individuals in this noncoding DNA can occur without apparent harmful effect (except for noncoding regulatory sequences). Thus, different individuals will show variations in the structure of their noncoding DNA, usually in the form of single base changes. Sometimes these changes occur in an existing recognition sequence for a restriction endonuclease and the site is then lost, or the change can create a new recognition site for a restriction endonuclease. When the DNA from an individual showing such a change is digested with the restriction endonuclease that cuts at that site, there will be a change in the size of the fragments produced compared with an individual not showing the change in base sequence. The DNA from the two individuals will therefore exhibit different fragment lengths when digested with the relevant restriction enzyme. These differences are RFLPs. Pieces of DNA can be inserted into the genome or deleted from it, by various mechanisms which need not concern us here, and these size changes can also be used to distinguish between individuals.

2.1.2 Southern blotting

To detect a specific piece of DNA within the total genomic background, for instance an RFLP or a gene with a particular mutation, the DNA must be available for manipulation and there must be a method for detection. One means of resolving DNA for manipulation is a Southern blot. DNA is cut into fragments of varying sizes using a restriction enzyme and the pieces are separated by electrophoresis in an agarose gel. Large fragments move more slowly and fragments between 500 bp and about 25 Kb can be separated with such a system. Agarose gels trap the DNA within them, but are very brittle, and cannot be easily manipulated, so the DNA is transferred by capillary action from the gel to a nylon membrane, where it is covalently bound. This is the blotting stage of the process, as shown in *Figure 2.1*. The DNA is now accessible on a solid support, arrayed by size, and specific sequences within the DNA can be analyzed. To examine the DNA, a 'probe' is used. This is a piece of DNA, RNA or synthetic oligonucleotide specific to the sequence of interest. It will bind only to that sequence, as long as the correct hybridization and washing steps are used. *Figure 2.1* shows how a radioactively labeled probe is detected. First, a radioactive nucleotide is incorporated into the probe, then the probe is hybridized with the nylon filter. In this process, conditions are such that the probe binds to the DNA on the filter, some to its specific target but the majority in a nonspecific manner. When hybridization is complete (this is usually an overnight step, at a temperature which depends on the hybridization fluid used), the

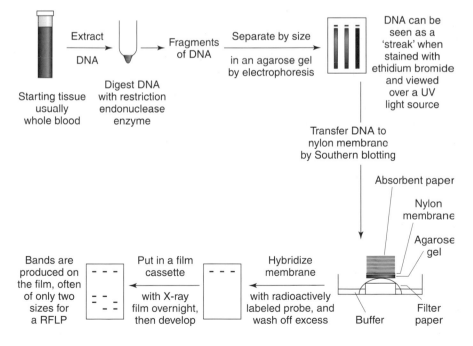

Figure 2.1: A schematic diagram showing the processes involved in Southern blotting. The end results are bands on X-ray film which correspond to restriction fragment length polymorphisms (RFLPs). (Reproduced from *Inherited Diseases of the Kidney*, Morgan and Grunford, 1998, by permission of Oxford University Press.)

nonspecifically bound probe can be washed off by using the correct salt concentrations in the buffer, leaving the probe bound only to its target sequence. As with 'melting temperature', mismatched sequences will separate at a different salt concentration compared to fully complementary sequences. The filter is then exposed to an X-ray film to produce an autoradiograph, which shows a band where the probe has hybridized. *Figure 2.2* shows the result of a Southern blot in a family with the genetic disorder Autosomal Dominant Polycystic Kidney Disease, and how such data can be used diagnostically in family is shown in section 3.2.1. *Table 2.1* shows the types of radiolabels commonly used in nucleic acid diagnosis, with some of their advantages and disadvantages. Probes can be labeled nonradioactively with a chemiluminescent fluorochrome, when the filter is treated as above or with a ligand that can be shown up in a colorimetric reaction on the filter itself.

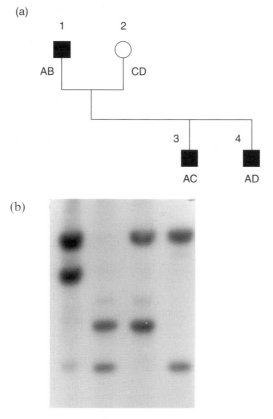

Figure 2.2: (a) Part of a family pedigree where adult polycystic kidney disease occurs due to a defect at the PKD1 locus on chromosome 16p13.3; (b) Autoradiograph of a Southern blot of DNA from each family member after digestion with endonuclease PvuII and hybridization with the highly polymorphic probe 3′HVR. This probe is on 16p13.3, and shows 5% recombination with the PKD1 locus. In this family allele *A* co-segregates with the disease.

Table 2.1: Types of radioactive labels used in nucleic acid diagnosis.

Label (half life)	Use	Advantages	Disadvantages
^{32}P (14 days)	Widely used in all types of analysis, e.g., Southern blotting, microsatellite analysis, ISH.	High specific activity, quick results, strong signal.	High energy beta emitter, more likely to cause damage. Bands on gels less clear. Does not give single cell resolution with ISH. Short half life.
^{33}P (25 days)	Can be used instead of ^{32}P.	Lower energy emitter than ^{32}P, therefore safer.	More expensive. Signal not as strong.
^{35}S (87days)	Sequencing gels and ISH.	High specific activity, low energy beta emitter, gives sharp sequencing bands and good localisation for ISH.	No disadvantages for uses given. Too low energy for Southern blotting.
^{3}H (12 years)	ISH	Low energy beta emitter that gives good tissue localization. Long half life.	Low activity probes. Long exposure times.
^{125}I (60 days)	ISH	Low energy gamma emitter. Good localization. Long half life	Long exposure times gamma emitter so Localisation worse than ^{3}H.

P, phosphorous, S, sulfur, H, hydrogen, I, iodine. ISH is explained in Section 2.5.1, and sequencing in Section 2.4.5. Microsatellites are detailed in Chapter 3.

For situations where size separation of fragments is not important, total DNA can sometimes be dropped in solution onto a nylon membrane and probed, this is known as a 'dot blot.' A variation on this theme is to use a manifold with slots in it connected to a vacuum pump to put the DNA or RNA onto a filter. This is called a 'slot blot.'

2.2 Types of probes

2.2.1 Oligonucleotides

There are a variety of nucleic acid molecules which can be used to probe DNA. The simplest are the synthetic oligonucleotides, which are usually in the region of 20–30 bp in length. These can be specific for particular mutations in genes when used in genetic analysis, or for regions of DNA that differ between strains of bacteria, in microbiology. Such probes are relatively easy to synthesize, and are robust in use.

2.2.2 DNA libraries and vectors

Longer probes are often derived from DNA 'libraries,' either genomic DNA (including all the noncoding and coding sequences) or cDNA (DNA complementary to messenger RNA) libraries which represent only the coding sequences found in particular tissues (*Figure 2.3*). Probes can also be made, commonly by PCR (see below). Libraries are grown in a variety of simple microorganisms, which have been extensively modified by molecular biologists. Such micro-organisms are termed 'vectors' and include plasmids, cosmids, phage and YACs (yeast artificial chromosomes). All the vectors share the common property that they can have foreign DNA inserted into their structure and will self replicate to produce copies of their own and the inserted DNA. Which vector is used depends on the size of the DNA to be inserted, and the requirements of the library (e.g. will it express the gene product, will a large number of copies of each insert be produced, see *Table 2.2*). cDNA libraries are made from mRNA, and therefore differ from tissue to tissue, genomic libraries are essentially the same from all normal tissues from the same type of organism. Some vectors, such as plasmids are much easier to work with than others, such as YACs. Plasmids are small, circular pieces of DNA which occur naturally in bacteria and carry the genes for antibiotic resistance. If a probe for an abundant mRNA, such as globin, was required, mRNA would be extracted from reticulocytes, made into cDNA using the enzyme reverse transcriptase, and inserted into the plasmid (*Figure 2.3*). The plasmids now have to be introduced into bacterial cells to be grown in large quantities. Bacteria are usually impermeable to large molecules such as long fragments of DNA. By treating the bacteria with high concentrations of calcium chloride, or by using electric 'shocks' the membranes can be made more permeable. This is called making the bacteria 'competent'. The bacteria will then take up the vector with its DNA insert ('transformation' of the bacteria). The cells usually take up a single vector, which will then replicate many times in each cell. Thus, each individual bacterium will have a single fragment of cDNA (or genomic DNA in a genomic library) within it – although in the case of cDNA libraries the most common mRNAs will be present in lots of cells, very rare ones could be missed altogether. To investigate a particular DNA molecule, individual bacteria must be grown into large cultures, the vector isolated, and the insert DNA removed. To obtain the correct insert DNA – what is 'correct' depends on the particular project – the library must be 'screened' (*Figure 2.4*). Screening involves using a molecule of some kind that will detect the insert DNA of interest. For example in cDNA libraries, if the equivalent cDNA from a related organism (e.g. mouse) has already been produced elsewhere, this can be radioactively labeled and used. Or, there may be an antibody to the human protein available which can be labeled and used if the library has been made in an 'expression

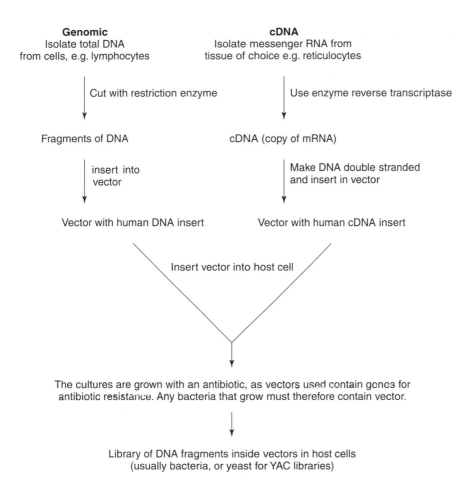

Figure 2.3: Making DNA libraries.

vector' (one which will produce a protein from its inserted DNA), although aberrant processing in bacteria can sometimes prevent this. Alternatively, part of the sequence of the DNA being investigated might be known, in which case a synthetic oligonucleotide can be made and labeled. In genomic libraries it might be that a particular region is being investigated, in which case a piece of DNA at the end of the existing known area could be used to screen the library to find the next piece (a process called 'chromosome walking' – see *Figure 2.5*). Whichever screening method is used, the end result is a bacterial 'clone' that contains a vector with the single insert of interest (*Figure 2.4*). Millions of copies of this can then be produced, the plasmid isolated, and the human cDNA cut out with restriction enzymes to use as a probe. The same procedure can be carried out for rare mRNAs but in this case a bacteriophage will often be used as a vector. Phage are viruses which infect bacteria very efficiently. They can take much larger inserts than plasmids, and they have a greater

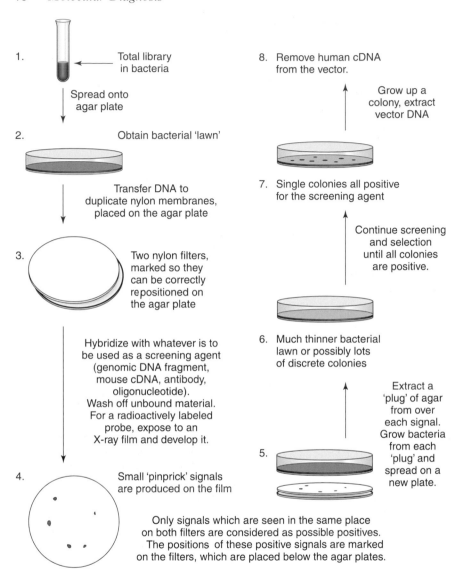

1.   Total library
      in bacteria

      Spread onto
      agar plate

2.   Obtain bacterial 'lawn'

      Transfer DNA to
      duplicate nylon membranes,
      placed on the agar plate

3.   Two nylon filters,
      marked so they
      can be correctly
      repositioned on
      the agar plate

      Hybridize with whatever is to
      be used as a screening agent
      (genomic DNA fragment,
      mouse cDNA, antibody,
      oligonucleotide).
      Wash off unbound material.
      For a radioactively labeled
      probe, expose to an
      X-ray film and develop it.

4.   Small 'pinprick' signals
      are produced on the film

8.   Remove human cDNA
      from the vector.

      Grow up a
      colony, extract
      vector DNA

7.   Single colonies all positive
      for the screening agent

      Continue screening
      and selection
      until all colonies
      are positive.

6.   Much thinner bacterial
      lawn or possibly lots
      of discrete colonies

      Extract a
      'plug' of agar
      from over
      each signal.
      Grow bacteria
      from each
      'plug' and
      spread on a
      new plate.

5.

      Only signals which are seen in the same place
      on both filters are considered as possible positives.
      The positions of these positive signals are marked
      on the filters, which are placed below the agar plates.

Figure 2.4: Screening a library.

'copy number', that is, they will produce a larger number of copies of the starting material than plasmids. For very large inserts, YACs can be used, as they will take inserts over 1 million bases (see *Table 2.2*). They are far more difficult to use, however. The principle of all libraries is the same: to insert human DNA from a complex mixture and isolate particular pieces of DNA to use as probes by screening. DNA from the vector can be labeled and used directly, or it can be used to produce an RNA copy – a 'riboprobe'. Certain vectors contain a promoter sequence for RNA polymerase, and the DNA insert acts as a template to produce RNA

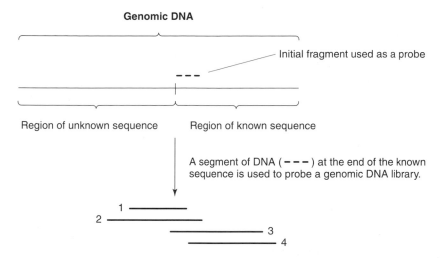

The probe will bind to fragments in the region already sequenced
(3 and 4), and in the region not yet sequenced (1 and 2). 3 and 4 are
discarded, while 1 and 2 are isolated and their ends used for more walking.

Figure 2.5: Chromosome walking.

Table 2.2: Vectors used in cloning

Vector	When used
Plasmid	If small fragments of DNA are to be cloned, or fairly abundant mRNA
Bacteriophage	Larger DNA fragments are needed (up to 50 kb), or very rare mRNA is to be cloned – since these vectors enter bacteria more efficiently than plasmids
P1 derived artificial chromosome (PAC) or bacterial artificial chromosome (BAC)	For cloning much larger DNA fragments – up to 200–300 kb
Yeast artificial chromosomes (YACs)	Can take insert DNA over 1Mb (1 million bases)

molecules which have ^{32}P or other radionuclides introduced as they are
made. These labeled molecules can be used in *in situ* hybridization
techniques, which are described later in the chapter (Section 2.5.1).

2.3 Polymerase chain reaction

The methods described above for separating DNA, and identifying particular
sequences, are still in use but by far the most common technique used in
nucleic acid analysis is the polymerase chain reaction (PCR). This is an
extraordinarily simple concept, for which Kary Mullis was awarded the
Nobel Prize for Chemistry in 1993, a fitting award for an invention that

has revolutionized the study of DNA and RNA. The basis of the method is that DNA polymerase will recognize a double stranded section of DNA with a single strand extending from it as a site to begin new strand synthesis. In practice, it means that a piece of DNA that was about one five millionth of the total genomic DNA, becomes the only visible component after gel electrophoresis and staining. Since the total amount of starting material is usually about 50 ng (10^{-9} gram), PCR produces a visible band on a gel from approximately 10^{-15} gram of DNA. Unlike Southern blotting, where the total genomic DNA can be a problem by masking the specific sequences, in PCR it is made irrelevant. The disadvantage of the technique is its very sensitivity, which can lead to false positives due to cross contamination.

2.3.1 Requirements for PCR and how it works

To use PCR, sequence information is needed for either side of the region to be studied. This is because synthetic oligonucleotides have to be made which are complementary to these areas and the positions of these 'primers' determines the limits of the piece of DNA which is to be amplified (see *Figure 2.6a*). A PCR reaction mixture contains about 50 ng of target DNA, specific primers, DNA polymerase, dATP, dCTP, dGTP, and dTTP to make the new strands, buffering agents, and $MgCl_2$. The target DNA is first made single stranded by heating the mixture to 95 °C, then cooled to allow the primers to bind.

When bound to the DNA, the primers are recognized by DNA polymerase as a site to begin synthesizing a new stand of DNA. For PCR, a thermostable DNA polymerase is used, often derived from the bacterium *Thermophilis aquaticus*, which grows in hot springs. This polymerase is known as *Taq* polymerase. Many of the available *Taq* polymerases are now cloned forms of the enzyme. In a PCR reaction there are usually between 20 and 40 cycles of amplification, carried out on a

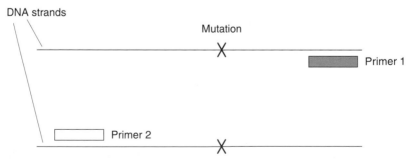

Figure 2.6a: Diagram showing DNA from a region of the genome where a mutation (X) is known to occur. This region is to be amplified by PCR. ▭ and ▨ are primers – synthetic sequences of DNA usually about 20 bases long – and the DNA fragment between them is that which is multiplied during the PCR reaction (see 2.6b and 2.6c).

machine designed for the purpose. Each cycle has three temperatures; 95 °C, 45 °C to 65 °C, and 72 °C (the middle temperature varies for different primers). The 95 °C step is to make the DNA single stranded, 45 °C to 65 °C allows the primers to bind (the annealing temperature) and 72 °C is the correct temperature for the *Taq* polymerase to make a new strand of DNA. However, often only 2 cycles are used, such as 60 °C and 72 °C, especially where short pieces of DNA are being amplified, as synthesis of a new strand can occur as the temperature moves from the lower to the higher.

Each cycle produces a doubling of the region of DNA to be amplified, so if all reactions worked at maximum efficiency 30 cycles would produce approximately 2 000 000 000 copies of each DNA fragment. In practice the yield is much less than this, and is dependent in part on the primer structure and cycling conditions.

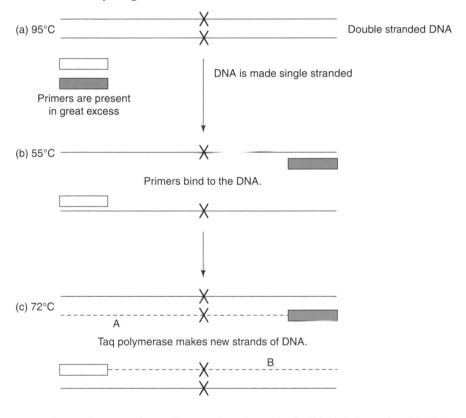

Figure 2.6b: First cycle in a PCR reaction. Step (a), the DNA is heated to 95 °C in the presence of excess primers, *Taq* polymerase, free nucleotides, $MgCl_2$ and buffer to maintain the correct pH. $MgCl_2$ is at 1.5 mM in standard PCR buffers, but must sometimes be adjusted to optimise the reactions depending on the primer base compositions. Step (b), the solution is cooled to around 55 °C (the exact temperature depends on the base composition of the primers), when the primers bind. Step (c), the reaction mixture is heated to 72 °C, the optimum temperature for *Taq* polymerase to extend the primers into a new strand of indeterminate length.

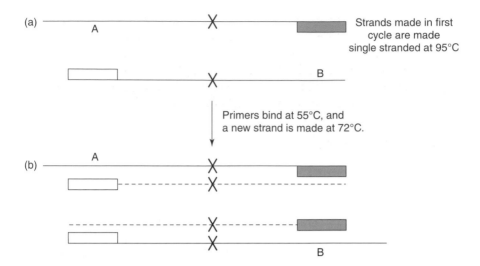

Figure 2.6c. Second and subsequent PCR cycles. (a) The DNA strands made in step (3) in fig. 2.6b, are separated at 95 °C. The new strand made using primer 1 (▭) has a binding site for primer 2 (▬) at position A, and the new strand made using primer 2 has a binding site for primer 1 at position B (see fig. 2.6b). (b) When the primers bind at A and B, and are extended by Taq polymerase, the DNA fragment can only go as far as the other primer, since there is no DNA past this to act as a template. Thus, in all subsequent cycles the primer positions limit the two ends of the fragments.

DNA strands are complementary in sequence, therefore each new strand synthesized will contain a binding site for the other primer in the reaction. In *Figure 2.6b*, the new strand made using primer 1 will have a site for primer 2 at position A. Likewise, primer 2 will form a strand with a site for primer 1 at position B. The first cycle in the reaction produces new DNA strands of indeterminate length but each subsequent cycle only makes DNA between the primer sites (*Figure 2.6c*).

Figure 2.6c shows how the position of the primers determines the size of the DNA fragment to be amplified. The site where the first primer binds in cycle 1 now limits the action of the Taq polymerase, as there is no DNA beyond it to act as a template.

When very low levels of target sequence may be present, for example in some case of viral detection, the sensitivity of PCR can be increased still further by using what are called 'nested' primers. For nested-PCR the first amplification is done as usual, then an aliquot of the solution is taken out and added to a second PCR reaction with a primer set (the 'nested' primers) that lie inside the sequence delineated by the original set of primers. A slightly smaller fragment than in the first round PCR is then produced, with a much greater lower limit of detection. This increased sensitivity is matched by a far greater likelihood of contamination, and stringent protocols with multiple negative controls are necessary for nested-PCR.

When PCR was first reported in 1985 there were no machines designed to cycle the temperatures, and *E. coli* DNA polymerase was used, which is not heat resistant. The reaction tubes had to be moved between water baths and new enzyme added at each cycle. It was the introduction of heat-resistant polymerase and computer controlled heating blocks (thermal cyclers) that made PCR a practical proposition.

2.3.2 RT-PCR and other modifications

PCR can also be used to examine RNA transcripts. This can be done in either two stages or one, depending on what is required. In the former case, the RNA is first converted to DNA using reverse transcriptase (RT) and the PCR reaction is performed on this cDNA template. For a single stage reaction, the enzyme *Tth* (from *Thermus thermophilus*) is used instead of *Taq* polymerase, as this enzyme has both DNA polymerase and RT activity. The involvement of a reverse transcriptase step in the process is the reason for it being known as RT-PCR. This method is used extensively for detecting viral RNA (see Chapter 4).

There are now several modifications to the basic PCR reaction which are used in diagnosis, such as amplification refractory mutation systems (ARMS) and these will be described in more detail under the relevant sections. PCR can also be used to amplify specific inserted DNA from vectors in genomic or cDNA libraries, using primers specific to the vector sequences.

2.4 Separation methods

There are numerous methods for separating DNA molecules, both those fractionated by enzyme digestion or the products of PCR reactions. Agarose gels are commonly used for both DNA (Southern blotting mentioned earlier) and RNA (Northern blotting). In the case of DNA, standard buffers are used and few precautions against degradation are required since DNA is very robust and DNAse enzymes are only likely to be a problem if gloves are not worn. For RNA more stringent precautions are required since RNAse is a ubiquitous enzyme. The RNA must be kept denatured while in the gel, and gels containing formaldehyde are most commonly used for electrophoresis of total RNA or mRNA. Solutions for RNA electrophoresis have to be treated with diethyl pyrocarbonate (DEPC), an RNAse inhibitor, as does all glassware used. The process of electrophoresis is similar for both Southern and Northern blotting, in that the agarose is dissolved in buffer by heating, then allowed to cool in a gel mould with a 'comb' to produce wells and when the gel is set it is submerged in buffer. Samples are placed in the wells, being kept in place with a dense loading buffer, and electrophoresed towards the anode.

For small DNA fragment separation, polyacrylamide gels are used instead of agarose (see Section 2.4.2).

2.4.1 Larger fragments

For DNA analysis of larger fragments, there are modifications of the electrophoretic process such as Field Inversion Gel Electrophoresis (FIGE), which will separate up to 150 000 bp and Pulsed Field Gel Electrophoresis (PFGE) where fragments as large as several megabases can be resolved. There are various designs of PFGE apparatus, but essential to all of them is a method of changing the direction of the current. Whereas ordinary electrophoresis applies current in a straight line, PFGE causes the current to change direction, at short intervals (the 'pulses'). This constantly changes the direction in which the DNA molecules are moving. The speed at which a molecule changes direction is dependent on its molecular weight, unlike conventional electrophoresis where large DNA molecules have a mobility independent of molecular weight. DNA is not extracted from whole blood as it is for Southern blotting, as this would shear the large fragments. Lymphocytes are separated from erythrocytes on a Ficoll gradient, removed by pipette and added to a solution of low melting point agarose. The mix is placed in small plastic moulds called 'block formers', in which the agarose sets into small blocks containing a known amount of cells. These cells are then ready for manipulation prior to PFGE (*Figure 2.7*).

The pretreatment given to the blocks of cells depends on the questions being asked. For a size separation, to find if two markers are physically near each other, the DNA in the block must be digested into fragments by

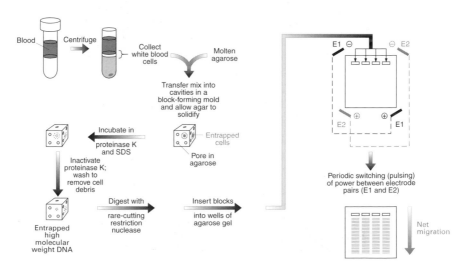

Figure 2.7: Fractionation of high molecular weight DNA from blood cells by pulsed-field gel electophoresis. (Reproduced from T. Strachan and A.P. Read (1996) *Human Molecular Genetics* 1st Edn. BIOS Scientific Publishers, Oxford.)

endonuclease enzymes. For PFGE, enzymes known as 'rare cutters' are used. These cut infrequently in the genome, either because they have a long recognition sequence, such as GGCCNNNNNGGCC (where N = any base) for the enzyme Sfi1, or they recognize GC rich sequences, that do not occur very often in DNA, for example, the enzyme Not1 (GCGGCCGC). For sizing of YACs the yeast chromosome plus the YAC are isolated in an agarose block, and run on the pulsed field gel (*Figure 2.8*).

Once the DNA has been digested, the blocks are sealed in wells in an agarose gel and the current is switched on for the desired time period, which can vary from 24 hours to several days depending on the size of the fragments under investigation. FIGE is a variation of PFGE, which uses a single electric field with periodic reversals of the polarity.

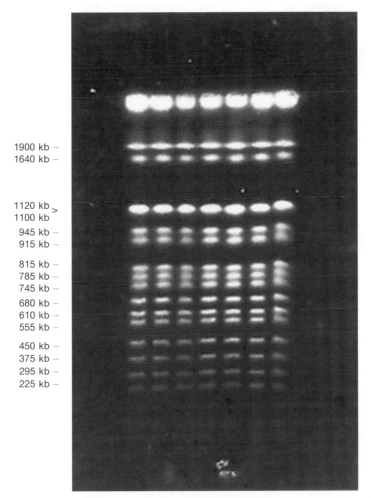

Figure 2.8: Yeast (*Sacchharomyces cerevisiae*) chromosomes separated by PFGE and visualized by ethidium bromide staining. Molecular weights are given by the side of the gel. Reproduced by kind permission of Amersham Pharmacia Biotech Limited.

2.4.2 Smaller fragments

DNA fragments between 100 and 20 000 bp can be separated on agarose gels. Different agarose concentrations are used depending on the size range to be fractionated, for example, for 100 bp pieces 3% agarose would be required, whereas for 20 000 bp 0.6% might be the concentration. There are modified agarose preparations which can be used to increase the sensitivity of separation in the lower size range. However, agarose is not sensitive enough to detect a one or two base pair difference between fragments. To achieve this, polyacrylamide gels are used. These separate in the range 10 to 1000 bp, and the concentration of polyacrylamide is altered depending on where the fragments being investigated lie within this range. Such gels are especially useful for small PCR products where there are only a few base pairs difference between individuals, for example, linked polymorphic markers in families with a genetic disorder (see Chapter 3).

2.4.3 Mutated DNA

Two techniques often used to find single base changes in PCR amplified DNA products are, single stranded conformational polymorphism (SSCP) analysis and denaturing gradient gel electrophoresis (DGGE). The first is more widely used, although it detects less mutations, because it is technically more simple. In both techniques, specific primers are used to amplify parts of the gene. SSCP utilizes the fact that single strands of amplified DNA take up a certain conformational structure when run on a polyacrylamide gel. If there is a missense mutation, this conformation can alter and a band of different mobility will be seen on electrophoresis. Double stranded DNA can also be used to detect mutations, using what is known as heteroduplex analysis (HA). In this case the PCR products are denatured and allowed to reanneal. If the normal copy of the gene is present only one type of double stranded DNA can be produced; a homoduplex. When a normal and a mutated copy of the gene occur, a mismatched double stranded fragment is formed from these two products, the heteroduplex, and this migrates with a different speed in a polyacrylamide gel. As with SSCP, the presence of the mutation leads to a change in the band pattern on the gel. It is possible to perform SSCP and HA on the same gel, although this leads to a loss of sensitivity for both methods, as the ideal electrophoresis times for the two products are usually different.

DGGE makes use of the changes in melting temperature of an amplified two stranded PCR product which can occur when a mutation is present. In this technique, denaturing gels with a gradient of urea are used, at raised temperatures. Amplified DNA is loaded on double stranded. This DNA migrates through the gel until it reaches a urea concentration where the strands separate and stop moving. Mutations will produce a different band pattern compared to the normal population.

2.4.4 Visualization

Double stranded DNA can be visualized on both agarose and polyacrylamide gels with the stain ethidium bromide (EtBr). DNA coupled with EtBr fluoresces bright orange under UV light (*Figure 2.8*). EtBr is a potent mutagen, however, and needs to be treated with care. Polyacrylamide gels can also be stained with silver nitrate, producing brown/black bands in normal light. If the PCR has incorporated a radioactive nucleotide, an autoradiograph of the gel can be produced as for Southern blotting.

2.4.5 Sequence analysis

Sometimes it is necessary to examine the actual nucleotide sequence of a length of DNA, and there are two main methods for performing such an analysis; the Maxam/Gilbert technique, and the Sanger sequencing procedure. The latter is by far the most commonly used now, and will be described here. The beauty of the Sanger process is its simplicity, which involves using dideoxy nucleotides in a reaction to produce new chains from the DNA template under investigation. These dideoxy molecules are incorporated into a new DNA molecule, but once there will not permit any additional dNTPs to be added to the chain since they lack the hydroxyl group required for attachment.

Sequences can be obtained from cloned DNA, or PCR products. The former are inserted into a phage vector called M13, which has a primer of known sequence adjacent to the site where the human DNA is inserted. M13 will synthesize single strands of DNA if a synthetic oligonucleotide complementary to the primer site is added, with DNA polymerase and the 4 dNTPs. The strands produced are released into solution. For sequencing, one dNTP is labeled, for example, with ^{32}P or a nonradioactive marker and four separate reactions are set up one with dideoxy ATP added to the mix, one with ddCTP one with ddGTP and one with ddTTP. The dideoxy nucleotides are randomly incorporated into the DNA strand produced from the template so that in the ddATP tube at every position an adenosine occurs on the chain, some product will be produced which ends at this point. The same is true for the particular ddNTP in other reaction tubes. When the reaction is complete, there are a number of different sized DNA molecules in each tube with a certain number terminating at every A, C, G, or T in the molecule. The four reactions are run out on a polyacrylamide sequencing gel which is exposed to an X-ray film if a radioactive label has been used. Each reaction produces a series of bands, like a ladder, on the film, and these can be

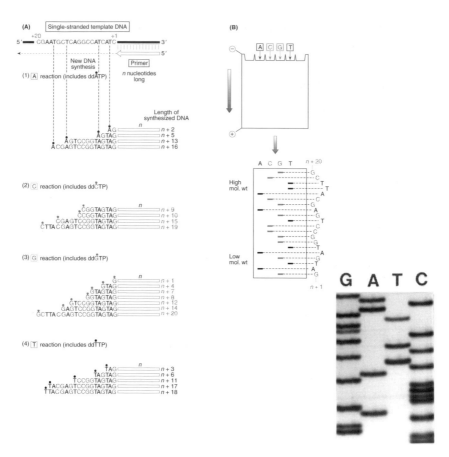

Figure 2.9: Dideoxy DNA sequencing relies on synthesizing new DNA strands from a single-stranded DNA template and random incoporation of a base-specific dideoxynucleotide to terminate chain synthesis. (a) Principle of dideoxy sequencing. The sequencing primer binds specifically to a region 3' of the desired DNA sequence and primes synthesis of a complementary DNA strand in the indicated direction. Four parallel base-specific reactions are carried out, each with all four dNTPS and with one ddNTP. Competition for incorporation into the growing DNA chain between a ddNTP and its normal dNTP analog results in a population of fragments of different lengths. The fragments will have a common 5' end (defined by the sequencing primer) but variable 3' ends, depending on where a dideoxynucleotide (shown with a filled circle above) has been inserted. For example, in the A-specific reaction chain, extension occurs until a ddA nucleotide (shown as A with a filled black circle above) is incorporated. This will lead to a population of DNA fragments of lengths $n+2$, $n+5$, $n+13$, $n+16$ nucleotides etc. (b) Conventional DNA sequencing. This generally involves using a radioactively labeled nucleotide and size-fractionation of the products of the four reactions in separate wells of a polyacrylamide gel. The dried gel is submitted to autoradiography, allowing the sequence of the complementary strand to be read (from bottom to top). The bottom panel illustrates a practical example, in this case a sequence within the gene for type II neurofibromatosis. (From T. Strachan and A.P. Read *Human Molecular Genetics* (1st Edn.) BIOS Scientific Publishers, Oxford)

'read' from the gel, giving the base sequence of the DNA. *Figure 2.9* summarizes this process.

2.4.6 Automated sequencing

There are now several automated DNA sequencing machines on the market which use fluorescence for rapid, high throughput sequence analysis. The Sanger method is still the principle behind the analysis, but M13 is not used. Instead, it is the ubiquitous PCR reaction which provides the samples for the sequencer. There are two ways of incorporating the fluorescence into the DNA to be sequenced, either by labeling the dideoxy nucleotides with a fluorescent dye, the dye terminator method, or by adding a fluorescent dye to the primers used in the PCR reaction, which is known as cycle sequencing. The samples are run in four lanes, as for manual sequencing, but the fluorescent fragments are 'read' by a laser in the sequencer. The bases are shown as a series of peaks on a paper printout (*Figure 2.10*). Sequencers can also be used to examine CA repeats (see Section 3.2), in which case the lanes have an internal size marker that allows the exact size of each frament to be given in basepairs. The sequencers are now often referred to as 'semi-automated' (see Section 6.4.2), as they have to be loaded manually, whereas the most recent robotic workstations will go from PCR to sequence in a fully auto-mated manner.

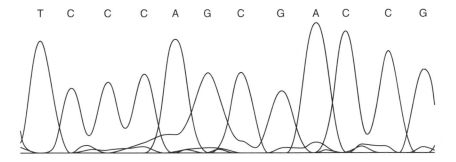

Figure 2.10: A sequencing trace from an ABI 377 semi automated sequencer. The bases have different colors on the trace; red for T, blue for C, green for A, and black for G.

2.5 Chromosome and tissue analysis

Nucleic acid diagnosis is not only concerned with examination at the molecular level, there are also various methods of tissue and chromosomal analysis which can provide information, some of which inevitably utilize the PCR reaction.

2.5.1 In situ hybridization (ISH)

Tissue sections or chromosome spreads can be examined with DNA or RNA probes, labeled with radioactivity or fluorescent dyes. For instance, to look at mRNA in a tissue section a radioactively labeled probe can be hybridized with the section, unbound probe washed off, and the radioactivity can be visualized by coating the slide with a liquid film. This film dries rapidly on the slide, which is then kept for days (or weeks) in the dark before the film is developed. Any bound probe can be seen as dark patches down the microscope. Riboprobes (Section 1.3) are especially sensitive for this type of work. Radioactive probes can also be bound to chromosomes. The commonest way to examine chromosomes is to look in metaphase, when the chromosomes are condensed, although there are now methods for interphase analysis. Cultured cells are induced to stop at metaphase by the use of colchicine, which inhibits formation of the 'spindle' that pulls the chromosomes apart. The position of a gene on a chromosome can be determined by binding a radioactively labeled probe to a metaphase spread (a slide with cells prepared on it that have been arrested in metaphase – see above).

In situ hybridization can also be used to detect viral particles (see Chapter 4). To increase the sensitivity of the technique, both for viral and chromosomal work, in situ can be combined with PCR, when it is called, not surprisingly, in situ PCR. This works by amplifying the sequence on the chromosomal spread or tissue section directly rather than extracting DNA or RNA as in a tube based PCR assay. RT-PCR in situ hybridization can be used to detect RNA viral particles, or cellular mRNA. The label, which can be radioactive, colorimetric or fluorescent, can be incorporated during the PCR reaction (direct in situ PCR), or added via a probe specific for the PCR product when the reaction is complete (indirect in situ PCR).

2.5.2 Fluorescent in situ hybridization, (FISH)

Chromosomal abnormalities used to be detected by Geimsa staining, which produces light and dark bands on the chromosomes. Changes in the sizes of these bands indicated a deletion of material, while enlargement suggested an insertion. Now, quite small deletions can be seen using fluorescent in situ hybridization (FISH). In this technique, the DNA probe is labeled by incorporation of modified nucleotides, obtained by binding of a 'reporter molecule' (e.g. biotin or digoxygenin), which can be detected by specific binding to another molecule. Following hybridization of the chromosome spread with the modified probe, excess probe is washed off, and a fluorescently labeled 'affinity molecule' is added which binds to the reporter. Under fluorescent microscopy a signal can be seen where the probe binds. A large probe such as a 40 kb cosmid is usually used for this technique to detect the signal, although plasmids or even small PCR fragments can be detected with computer enhancement of the image.

An extension of this technique is chromosome painting, where DNA from individual chromosomes is used as a fluorescent probe. By using different mixes of fluorochromes in each chromosomal DNA mixture, it is possible to produce a different color for each chromosome. This is particularly useful when examining chromosomal rearrangements (see Section 3.5).

2.5.3 Comparative genome hybridization (CGH)

This specialized use of FISH is used in tumor analysis to identify gains and losses of chromosomal material (see Section 6.4.4). DNA from a tumor and from a normal control individual are labeled with different colored fluorescent labels and hybridized together to normal metaphase spreads. An abnormal ratio in signal intensity between the two signals indicates a region of loss or gain of material.

2.5.4 Protein truncation test

The above techniques all involve examination of nucleic acids directly in some way, but there is also a method of detecting mutations in some genetic disorders by examining the protein product of a specific gene. This is the protein truncation test (PTT) and is used when most mutations produce a truncated protein from the mutated gene, that is when stop codons are generated. The test can be done from genomic DNA, but usually RNA is the starting material. *Figure 2.11* summarizes the process. First, reverse transcriptase is used to make a cDNA copy of the RNA. Primers for PCR are made to amplify the area of interest of the gene, and one primer also contains a promoter for RNA polymerase and a translation initiation sequence at its 5′ end (PCR primers can have extensive mismatches at the 5′ end without affecting the reaction, as bases are added to the 3′ end only). The amplified DNA from the PCR reaction is used in a reticulocyte lysate system that couples transcription and translation. Radioactive amino acids are used in the reaction, and the reaction products are run on an acrylamide gel, which is then exposed to an x-ray film. The resulting autoradiograph has one band for each sample producing normal sized product, plus a smaller band where a mutation gives rise to a truncated product (*Figure 2.11*).

2.6 Some pitfalls in nucleic acid diagnosis

As with all laboratory techniques, good practice is essential in nucleic acid diagnosis. This is especially true of PCR where the sensitivity of the technique can be a real difficulty. For diagnosis of genetic diseases there is

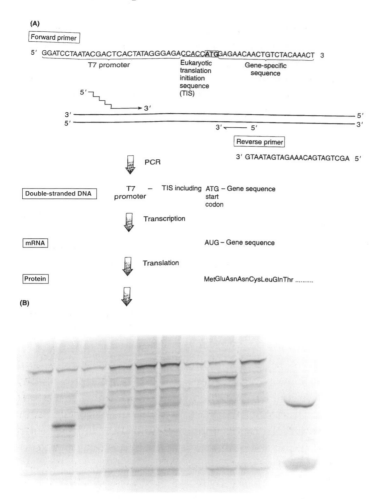

Figure 2.11: The protein truncation test (PTT). Coding sequence without introns (cDNA or large exons in genomic DNA) is PCR amplified using a special forward primer that includes a T7 promoter, a eukaryotic translation initiator with an ATG start codon, and a gene-specific 3′ sequence designed so that the sequence amplified reads in-frame from the ATG. A coupled transcription–translation system is used to produce polypeptide from the amplified coding sequence, which is then checked for size by gel electrophoresis. A truncated polypeptide points to the presence of a stop codon in the amplified coding sequence. In this example, the primers were designed to amplify a 2-kb region in exon 15 of the APC (adenomatous polyposis coli) gene, using a modification of the method of van der Luijt *et al.* (1994). The bottom panel shows translation products resolved by SDS-polyacrylmide gel electrophoresis (PAGE). Lanes from left to right are: 1–9, unrelated familial adenomatous polyposis coli patients; 10, negative control; 11, size markers (46 kDA and 30 kDA). In addition to the wild-type translation product of 72 kDA (top band), note the presence of truncated products in lanes 2 (41 kDA), 3 (50 kDA) and 8 (64 kDA). Original photograph kindly provided by Tayebeh Hamzehloei, University of Newcastle upon Tyne. (Reproduced from T. Strachan and A.P. Read (1996) *Human Molecular Genetics* 1st Edn. BIOS Scientific Publishers, Oxford.)

not usually a great problem, as there is so much target DNA in each sample. Even so, great care has to be taken, as it is the PCR product, rather than DNA from other individuals, that is the most likely cause of contamination. These short stretches of DNA are preferentially amplified as they do not have the complex structure of total genomic DNA. If contamination does occur it can usually be eliminated by renewing all solutions.

For detection of viral nucleic acids, the problem is much greater. In this case the target can be present in very low copy number, often at the limits of sensitivity of PCR, and just a few copies of PCR product can prove disastrous. Also, it is not possible to check the parents of a virus to see if the result makes genetic sense! To minimize the problem, solutions for PCR are prepared in one room, sample is added in another, and the PCR is carried out in another. PCR products are never allowed into the other areas. Some laboratories have different colored lab coats for those people in each room, and different colored pipettes, so it is immediately obvious if anyone or anything is in the wrong place. This may seem excessive, but there are many papers in the literature where positive viral results have proved to be due to contamination, and this has also occurred in archeological specimens (see Chapter 7).

Southern blotting does not suffer from the sensitivity of PCR, but there are associated difficulties. Sometimes there can be faults in the membrane, so that a band which should be there does not show up. If restriction enzymes are used on total human DNA bands may occur where they are not expected due to 'partial' digestion. This simply means that not all the DNA has been properly cut by the enzyme, usually producing larger bands than expected as well as those of the correct size. To guard against this, the digested DNA is examined on the gel by ultraviolet light, with EtBr in the gel. A good digest gives an even fluorescent stripe on the gel, a partial digest is usually brighter at the top.

Mutations in the primer sites in PCR can produce anomalous results, especially if they are at the most 3′ position. In this case the primer will almost certainly not amplify, and this can give the appearance of a deletion for this region, whereas it is simply a point mutation. These 'null alleles' are not common, but they have to be considered.

With all mutation detection systems the problem arises of when is a change in base sequence causative of a problem and when is it simply a neutral change? This is particularly true in some genetic disorders where changes in the gene may be reported in single families, or even isolated cases. If the change in amino acid is very nonconservative (i.e., the amino acid produced from the mutated codon is markedly different in size and/or charge) this is suggestive of a disease causing mutation, and if a large number of unaffected individuals do not show the change this also adds weight to the proposition. However, neither occurrence makes it certain that the change is producing the disease, and there are several reports of mutations that were considered causative, and fulfilled both previously mentioned criteria, being later shown to be benign.

Lastly, much DNA based technology uses toxic reagents, although there is always movement towards producing safer methodologies, for example, radioactive labels are replaced whenever possible by colorimetric assays or silver staining (but even these may be toxic). The methodology can be expensive if commercial kits are used, but even 'in house' methods are not necessarily cheap. Despite its possible pitfalls, nucleic acid based technology has produced great changes in the diagnosis of disease, and will undoubtedly continue to do so.

2.7 Future diagnostic research

Nucleic acid diagnosis is at present quite labor intensive, although there are automated sequencers and robots to perform PCR reactions that require large numbers of the same reaction. The use of robotics is certain to increase in the next few years as DNA testing becomes more common. The area where technology will have the greatest impact will be the use of silicon chips as the matrix for oligonucleotide probes that detect mutations. In this method, short oligonucleotides (about 15 bases long) are attached to a silicon chip, and this is used as a hybridization target for an individual's DNA. At present the technique is being applied to known mutations in known genes, and has already been tried with success for mutations in one of the breast cancer causing genes, *BRCA1*.

Despite remaining technical difficulties, chips for known mutations in many genes cannot be far away. The final aim will be a chip, or series of chips, that covers all known genes, and can detect any mutation, whether known or not. This may appear firmly in the realms of science fiction, but genetics and technology are moving so fast that it might not be as far away as now seems likely. Given the probability that such advances will occur, their application ought to be a debate for the present. A historical analysis of how new technologies impacted on human society does not give much grounds for optimism in the field of genetic analysis. For this pessimism to be proved unfounded, the development of society will have to have exceeded humanity's undoubted technical ingenuity.

This chapter has outlined many methods used in nucleic acid based diagnosis. In the following chapters, the way these methods are used in particular cases will be examined.

Further reading

Botstein, D., White, R.L., Skolnick, M. and Davis, R.W. (1980) Construction of a genetic linkage map in man using restriction fragment length polymorphism. *J. Hum. Genet.* **32:** 314–331.

Davies, K.E. (ed.) (1988) *Genome analysis, a practical approach*. IRL Press.

Elles R. (ed.) (1996) *Molecular Diagnosis of Genetic Disease*. Humana Press.

Hiyashi, K. (1991) PCR-SSCP: a simple and sensitive method for the detection of mutations in genomic DNA. *PCR Meth. Appl.* **1:** 34–38.

Maxam, A. and Gilbert, W., (1977) A new method for sequencing DNA. *Proc. Natl. Acad. Sci. USA* **74:** 560–564.

Myers, R.M., Fischer, S.G., Lerman, L.S. and Maniatis, T. (1987) Detection and localisation of single base changes by denaturing gradient gel electrophoresis. *Methods in Enzymol.* **155:** 501–507.

Saiki, R.K., Scharf, S., Faloona, F., Mullis, K.B., Horn, G.T. and Erlich, H.A., et al. (1985) Enzymatic amplification of beta-globin genomic sequences and restriction site analysis for diagnosis of sickle cell anaemia. *Science* **230:** 1350–1354.

Sambrook, J., Fritsch, E.F. and Maniatis, T. (1989) *Molecular Cloning: A Laboratory Manual.* Cold Spring Harbor Laboratory Press.

Sanger, F., Nicklen, S. and Coulston, A.R. (1977) DNA sequencing with chain-terminating inhibitors. *Proc. Natl. Acad. Sci. USA* **74:** 5463–5467.

Southern, E.M. (1975) Detection of specific sequences among DNA fragments separated by gel electrophoresis. *J. Mol. Biol.* **98:** 503–508.

Spurr, N.K., Young, B.D. and Bryant S.P. (eds.) (1998) *ICRF Handbook of Genome Analysis.* Blackwell Science.

Van der Luit, R., Khan, M., Vasen, H., van Leeuwen, C., Tops, C., Roest, P., den Dunnen, J. and Fodde, R. (1994) Rapid detection of translation-terminating mutations at the adenomatous polyposis coli (APC) gene by direct protein truncation test. *Genomics* **20:** 1–4.

van Prooijen-Knegt, A.C., van Hoek, J.F.M. and Bauman, J.G.G., et al. (1982) In situ hybridisation of DNA sequences in human metaphase chromosomes visualised by an indirect fluorescent immunocytochemical procedure. *Exp. Cell Res.* **141:** 397–407.

Chapter 3

Human genetic diseases

3.1 Types of genetic disease and patterns of inheritance

Human genetic diseases fall into two main categories:

(i) Single gene disorders, where mutations in one gene produce the disease, for example, cystic fibrosis and Duchenne muscular dystrophy.
(ii) Polygenic disorders. These diseases have a genetic component, which may be due to a combination of genes, or genes and environmental factors, for example, hypertension and diabetes.

Single gene disorders are the best studied, because the research is more straightforward, with the result that, for diagnostic purposes, there is more testing performed for these diseases. Polygenic, multifactorial diseases represent the future direction of genetic research, and eventually, diagnosis.

There are far too many single gene disorders now being tested to examine them all in this chapter. A range of disorders which illustrate most problems faced in the genetic diagnostic laboratory will therefore be discussed.

Single gene disorders show three patterns of inheritance, and it is very important to know how the gene is being passed through the family, especially if linkage analysis to an unknown gene in a known chromosomal position is to be carried out (see Section 3.3). Diseases which affect males and females equally are carried on the autosomes (chromosomes 1–22), whereas 'sex-linked' diseases are transmitted on the X chromosome. Both categories of disease show either 'dominant' or 'recessive' patterns of inheritance.

3.1.1 Autosomal dominant

In these cases, only one mutated copy of the gene is required to produce the disease. When drawn, the pattern of inheritance is said to be vertical (see *Figure 3.1a*). There can be father to son transmission, and males and females are equally affected, for example Huntington's Chorea, autosomal dominant polycystic kidney disease, neurofibromatosis, myotonic dystrophy.

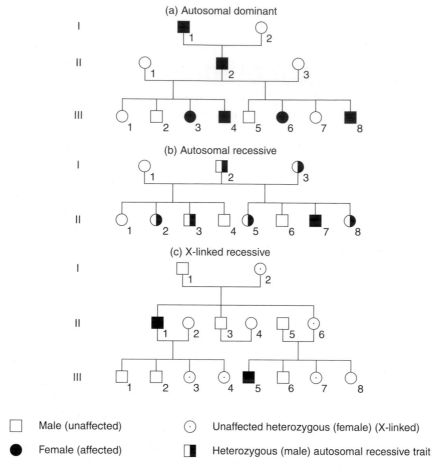

Figure 3.1: (a) Typical examples of idealized patterns of inheritance for autosomal dominant disease; (b) autosomal recessive disease; and (c) X-linked disease. (Reproduced from *Inherited Diseases of the Kidney*, Morgan and Grunfeld, 1998, by permission of Oxford University Press.)

3.1.2 *Autosomal recessive*

In these disorders both copies of the mutated gene must be present for the disease to occur. The pedigrees show a 'horizontal' pattern of inheritance (see *Figure 3.1b*), with father to son inheritance possible. Examples include cystic fibrosis, and sickle cell disease. The incidence of these disorders is increased in areas where marriage between familial relatives, such as first cousins, is common. In such 'consanguineous' marriages it is more likely that each partner will carry a mutated copy of a disease causing gene, if it exists within the family, and therefore have affected offspring.

3.1.3 X-linked disorders

X-linked dominant diseases are very rare, vitamin-D resistant rickets being an example. For diagnostic purposes this mode of inheritance can be ignored. X-linked recessive diseases are important and include hemophilia and Duchenne and Becker muscular dystrophies.

Females carry the disease, but *very* rarely express it, due to the presence of another copy of the gene on the second X chromosome. There is no father to son transmission (*Figure 3.1c*), since no X chromosome is transmitted.

Tests for single gene disorders include looking for known mutations in a well characterized gene, looking for unknown mutations in a known gene, or investigating a family where only the chromosomal position of the gene responsible for the disease is known. The last two cases both rely on using linkage between the gene and a nearby, polymorphic marker (Section 3.3).

3.2 Markers used in linkage analysis

The principle of linkage analysis is given in Section 3.3., but one requirement is a marker near the gene under investigation. Such markers must vary in size in different individuals, and the greater the variation the better, that is the markers must be highly polymorphic in the population. Some years ago, the major class of markers were RFLPs (Section 2.1.2), but these have been superseded by CA repeats (also called microsatellites). These are repeat stretches of cytidine and adenosine, scattered throughout the human genome in noncoding regions, with no known function (which is not the same as no function!). These repeats are easy to amplify by PCR (see Section 2.3), and are highly variable in the population. Primers are made complementary to the DNA either side of the CA repeats, and the PCR products are separated on polyacrylamide gels. Visualization can be by staining, or by autoradiography if the PCR has included radioactive components. *Figure 3.2* shows a silver stain of a CA repeat, with details of the family members from which the DNA was obtained. The CA repeat bands seen on a gel can be sized by the number of base pairs if automated sequencing is used (see Section 2.4.6), or they can be numbered from the top of the gel downwards, as indicated in *Figure 3.3*. In this case CA repeat size cannot be compared among different families, as band 1 in one family will not be the same as band 1 in another family. For linkage analysis this does not matter, as it can only be done **within families**, but there are other analyses outside the scope of this book where this lack of data can be important.

In some cases RFLPs are used, and where possible these have been converted to a PCR format by sequencing around the restriction enzyme sites.

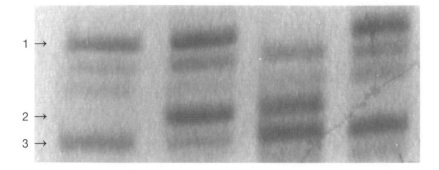

Figure 3.2: An example of micro-satellites amplified by PCR in a family with adult polycystic kidney disease (part of a much larger pedigree). The micro-satellite in this case is SM7 which shows about 1% recombination with the PKD1 locus. The first lane is DNA from the affected mother, lane 2 is from the unaffected father, then lanes 3 and 4 are affected children. The disease in this family is caused by a mutation at the PKD1 locus. The two darker bands for each individual represent the alleles, the fainter bands are so-called stutter bands. There are three alleles in this family (arrowed). The mother has alleles 1 and 3, the father 1 and 2. Each child has inherited allele 3 from the mother, which in this family is a marker for the disease gene. (Reproduced from *Inherited Disorders of the Kidney*, Morgan and Grunfeld, 1998, by permission of Oxford University Press.)

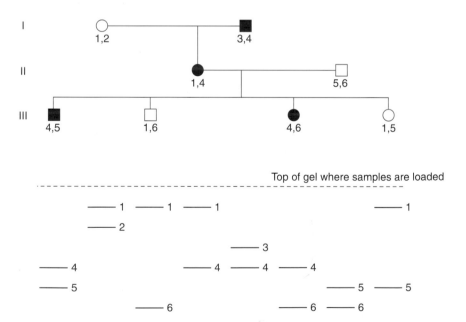

Figure 3.3: Family pedigree showing an autosomal dominant disease with marker data from a CA repeat. The bands for the CA repeat, as they would be seen for each individual on a gel, are shown in the schematic below the pedigree.

3.2.1 Family analysis using linked markers

There are three reasons for using linkage analysis in a family where a genetic disease is known to occur.

(i) Where the chromosomal location, or 'locus', is known, but the gene itself has not yet been isolated (now a small number in diagnostic terms, for example, tuberous sclerosis type 1 on chromosome 9q34).

(ii) Where the gene is known but mutations are very varied and hard to find (e.g. autosomal dominant polycystic kidney disease type 1).

(iii) Where common mutations can be detected easily, but rarer ones are very time consuming (e.g. cystic fibrosis).

In all these instances, potentially affected individuals or carriers have to be identified by using nearby markers known to be 'linked' to the gene. A particular size of marker therefore indicates the position of the gene causing the disease. Such analysis can only be used in individual families, not in the population as a whole, and the concept is a simple one. A marker, which is known to be close to the disease locus, is analysed in each family member. This is usually a CA repeat. The marker data is put onto the family pedigree (*Figure 3.3*), and it can be seen that in the case shown, all affected individuals have band 4, while none of the unaffecteds show this band. In this family, persons with band 4 will develop the disease, while those without this marker band will not, within the error limits of the technique. This last point is important, and will be expanded on below.

Firstly, it should be pointed out that *Figure 3.3* represents an idealized, and rather unlikely situation, where both grandparents have completely different band sizes, and their son-in-law has different band sizes again. It is much more likely that some sizes of CA repeat are more frequent in the population than others, and will therefore be over represented in families. For diagnostic purposes, the closest, most 'informative' marker must be used. If it is not informative, another marker must be used. *Figure 3.4* illustrates the concept of informativity. *Figure 3.4a* shows a completely uninformative marker, where both chromosomes in each individual are the same, and this would be diagnostically useless. There is a little more information in 3.4b, as the unaffected mother has two sizes of marker, but this as the affected father is still homozygous the problem is the same as in 3.4a. *Figure 3.4c* shows a situation which is informative, where the affected person has 2 different sizes of CA repeat, and band size 2 in this case tracks with the affected gene in this family. The affected daughter's sister is therefore not affected, within the error rate of the diagnosis. When RFLPs were the predominantly used markers, informativity was often a problem, as many RFLPs show only 2 sizes in the population. For CA repeats there is usually a marker in the desired region that will prove informative.

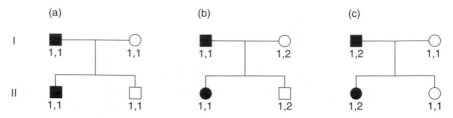

Figure 3.4: Three pedigrees illustrating informativity of markers. The numbers under each individual represent CA repeat sizes as seen as bands on a gel, as indicated in *Figure 3.3*.

3.2.2 Error rate using linked markers

If the marker was always inherited with the disease gene, there would be no error in the diagnosis (except human error), but this is not the case. At meiosis, when the germ cells are formed (sperm or egg), 'recombination' or 'crossing-over' can occur between the CA repeat and the gene (*Figure 3.5*). The likelihood of this recombination occurring is the error of the

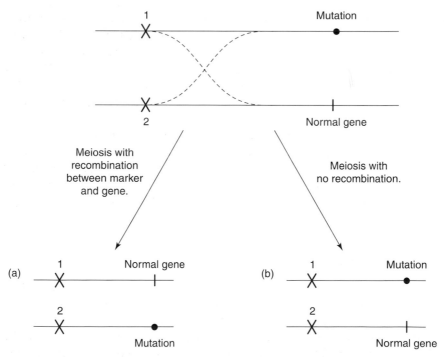

Figure 3.5: Diagram to illustrate what happens to a gene and a nearby marker at meiosis, (a) with no recombination, and (b) with recombination. '1' and '2' represent different marker sizes, such as CA repeats, on each chromosome. The dotted line represents what happens if there is a recombination at meiosis (situation 'a').

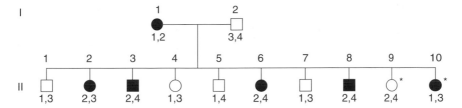

Figure 3.6: A pedigree with an autosomal dominant disorder, showing inheritance of a marker and disease. The two individuals asterisked (*) in generaion II show recombination between gene and marker. The numbers above each individual are for identification, those below represent marker bands on a gel (see *Figure 3.3*).

diagnosis, and is roughly dependant on the distance between the marker and the gene. The smaller the distance, the less likely recombination is to occur. To ascertain how often recombination occurs between each marker and a disease gene – known as the 'recombination fraction' – a series of families with the genetic disease have to be tested with each marker, and the number of recombinants and non-recombinants scored. In *Figure 3.6*, there are 2 recombinants. Individual II_{10} is affected, but has inherited marker 1 from his mother, while all other affected persons have inherited marker 2, while II_9 is unaffected, but has inherited marker 2. In this family, the number of recombinants, 2, divided by (the number of non recombinants + the number of recombinants), 10, is 2/10 or 20%, or 0.2. The recombination fraction gives the error of diagnosis, in this case 20%. For diagnostic purposes, the smaller the recombination fraction the better.

For diagnostic use, probes have already been evaluated for their degree of polymorphism and amount of recombination with a given disease locus by research laboratories. The question of whether a particular probe is or is not linked to the gene therefore does not have to be considered. Those interested in this concept should consult *Human Molecular Genetics*, by Strachan and Read, Chapter 12 (BIOS Scientific Publishers, Oxford).

3.3 DNA for prenatal diagnosis

The following sections will detail how different mutations can be detected, and these can be for presymptomatic analysis or prenatal diagnosis. In the latter case it is necessary to get DNA from the fetus. This is done by taking a needle biopsy of material from the trophoblast layer of the placenta, which is fetal in origin. This can be done either by intra-uterine or transabdominal sampling, and is usually carried out at 8–12 weeks. The sample obtained is called a chorionic villus sample, or CVS, and it is examined under a microscope for maternal contamination before the DNA is extracted. The removal of a CVS has an associated risk of miscarriage, but this is now very low, at 1%.

3.4 Detecting known mutations

Mutations that cause genetic disease are many and varied. For the purposes of this chapter we will divide them into five groups, and give examples of diagnostic methods used for each type. The groups are (1) point mutations, (2) deletions, small and large, (3) triplet repeat expansions, (4) rearrangements and duplications, and (5) imprinting.

3.4.1 Point mutations

Single base changes are the most common mutation, and can lead to no amino acid change (due to the degeneracy of the DNA code), a conservative amino acid change that produces no phenotypic change, such as glycine to alanine, a nonconservative amino acid change which causes an effect which may lead to a disease, or a premature stop codon which produces a truncated protein (which is often degraded). Also, the existing stop codon can be changed to code for an amino-acid, in which case extra amino acids are added to the end of the normal protein. How many are added depends on when the next stop codon happens to lie in the 3′ end of the gene.

An example of a deleterious base change which alters an amino acid is sickle cell anemia. This is an A to T change, altering the CAG codon for glutamic acid to the CTG codon for valine in the B-globin gene on chromosome 11 (*Figure 3.7*). This mutation is found only in people who originate from areas where malaria is prevalent, and is absent from caucasians. This is because the heterozygous state, where only one copy of the gene is mutated, affords some protection against infection by the malarial parasite. Unfortunately, the homozygous condition (where both genes carry the base change) causes significant mortality due to blockage of the capillaries under conditions of low oxygen, when the altered hemoglobin tends to come out of solution. Improved clinical care has greatly extended the life span of affected individuals, but the disease still produces both childhood and adult deaths. Unusually, this coding change also affects a restriction enzyme cutting site, in this case for Mst II (*Figure 3.7*), and this means that RFLPs and Southern blotting can be used to detect the mutation, by using part of the beta-globin cDNA as a probe. The mutation abolishes an MstII site, with the result that the sickle cell gene (β^s) gives a 1.4 kb band on the X-ray film, while the normal gene (β^A) gives a 1.2 kb band (*Figure 3.7*). In this case there is a no recombination error, as the RFLP detects the mutation itself. As with most tests, this is now converted to PCR, when a small fragment around the restriction site is amplified, and the PCR product cut with MstII or another enzyme that recognises the same cutting site.

ARMS
Detection of a point mutation can be by direct amplification of the DNA region containing that mutation and use of a restriction enzyme where

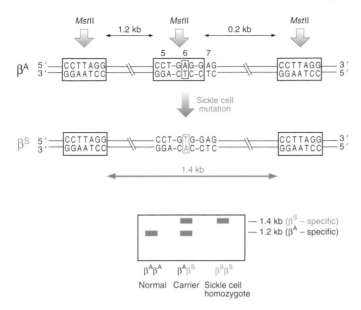

Figure 3.7: Diagram showing the point mutation in the beta-globin gene responsible for sickle cell disease. The bands seen on a southern blot using a probe for part of the beta-globin gene are shown in the rectangle at the bottom of the figure; 1.2 kb for the normal gene and 1.4 kb for the mutated gene. (Reproduced from T. Strachan and A.P. Read (1996) *Human Molecular Genetics* 1st Edn. BIOS Scientific Publishers, Oxford.)

appropriate, as described above for sickle cell anemia. An alternative is to use another PCR based approach known as ARMS. This can be used for any point mutation. ARMS utilizes the fact that PCR primers must be complementary at the 3' ends. It is also referred to as allele specific amplification (see Section 6.4.2).

A primer is made complementary to the normal gene at the 3' end, and one complementary to the mutant gene, with a common primer to complete the amplification (*Figure 3.8*). The specific primers also have the same mismatch of another base about 4 bases from the 3' end. This is because PCR can sometimes tolerate one mismatch, but not two.

For each individual, two PCR reactions would be set up, one with the mutant specific primer, and one with the normal specific. An individual homozygous for the mutation would only produce a product in the tube with the mutant specific primer, while an individual with only the normal gene would give a product with the normal specific primer. A person with one copy of each (a heterozygote) would produce a product in each tube. To ensure that an absence of signal is not due to PCR failure, a primer pair not connected with the mutation, which produces a different sized band, is used in each tube.

ARMS primers for different mutations in the same gene can be designed to give different sized products, and can be amplified together, a

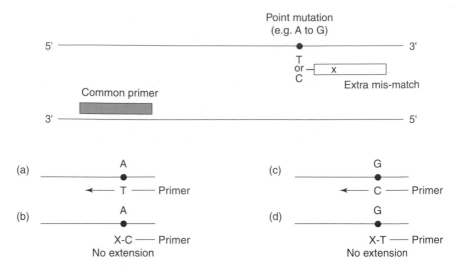

Figure 3.8: ARMS primers for an A to G point mutation. The primer [X   ] has a T at the 3' end for the non-mutated gene, and a C for the mutated gene. There is a built in mismatch (X) to ensure that PCR will not occur if the 3' base is not complementary. The primer [▓▓▓▓▓▓] is common to both reactions. Figures 3.8a–d show which situations would allow amplification and which would not for the two different bases in the gene with the two possible primers.

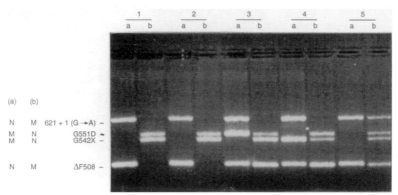

Figure 3.9: Multiplex ARMS test to detect four cystic fibrosis mutations. The common primers have been designed so that each reaction gives a product of a different size, allowing multiplexing. Each tube contains primers for two normal and two mutant sequences. PCR products, from top to bottom, test for the mutation $621+1$ (G→A), G551D, G542X and ΔF508. In the left track of each pair the ARMS primers amplify the normal counterparts of the $621+1$ (G→A) and ΔF508 sequences, and the mutants G551D and G542X; in the right hand track the primers amplify the opposite allele in each case. Tracks 1,2: no mutation detected; track 3: compound heterozygote for ΔF508 and G551D; track 4: compound heterozygote for ΔF508 and G542X; track 5: compound heterozygote for ΔF508 and $621+1$ (G→A). Courtesy of Dr Andrew Wallace, St Mary's Hospital, Manchester; data obtained using a kit supplied by Cellmark Diagnostics, Abingdon, UK. (Reproduced from T. Strachan and A.P. Read (1996) *Human Molecular Genetics* 1st Edn. BIOS Scientific Publishers, Oxford.)

process termed multiplexing. *Figure 3.9* shows a multiplex ARMS reaction for four mutations in the cystic fibrosis gene.

Allele specific oligonucleotide analysis

Point mutations can also be detected with allele specific oligonucleotides (ASOs). These are synthetic sequences of nucleotides, usually about 20 bases long, one of which is complementary to the region where the point mutation occurs, the other of which is complementary to the same region in the nonmutated gene. Sickle cell disease can be detected in this way, as shown in *Figure 3.10*. The region of the gene where the mutation occurs is

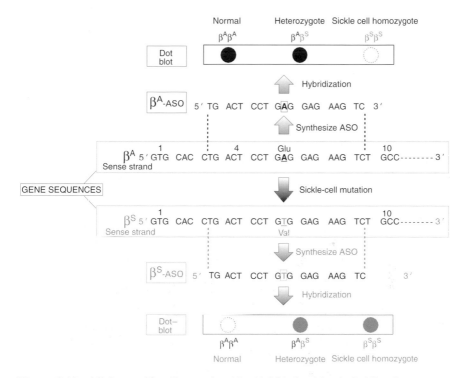

Figure 3.10: Allele-specific oligonucleotide (ASO) dot-blot hybridization can identify individuals with the sickle cell mutation. The sickle cell mutation is a single nucleotide substitution (A→T) at codon 6 in the β-globin gene, resulting in a GAC (Glu)→GTG (Val) substitution. The example shows how one can design ASOs: one specific for the normal (β^A) allele and identical to a sequence of 19 nucleotides encompassing codons 3–9 of this allele, and one specific for the mutant (β^S) allele, being identical to the equivalent sequence of the mutant allele. The labeled ASOs can be individually hybridized to denatured genomic DNA samples on dot-blots. The β^A- and β^S-specific ASOs can hybridize to the complementary antisense strand of the normal and mutant alleles respectively, forming perfect 19-bp duplexes. However, duplexes between the β^A-specific ASO and the β^S allele, or between the β^S-specific ASO and the β^A allele have a single mismatch and the ASOs are removed with correct washing conditions. (Reproduced from T. Strachan and A.P. Read (1996) *Human Molecular Genetics* 1st Edn. BIOS Scientific Publishers, Oxford.)

amplified by PCR. The products are either run on an agarose gel, which is Southern blotted using a membrane on each side, or the products are dot blotted onto two membranes (see Section 2.1.3). Although the former method takes longer, any signal obtained can be checked to ensure it is the same size as the PCR products, which is not the case for a dot blot. Radioactive ^{32}P-CTP is incorporated into each of the ASOs and these are hybridized, one with each membrane. By using the correct washing conditions the oligonucleotide for the point mutation will only bind to that sequence, while the ASO for the nonmutated sequence will only bind there. After washing, the membranes are exposed to x-ray film to produce an autoradiograph (see Chapter 1, Section 1.1.3). Homozygotes for sickle cell only give a signal (a band after a southern blot, a dot after a dot blot) with the ASO for that mutation, homozygotes for the nonmutated sequence only give a signal with that ASO, while heterozygotes give signals in both cases (*Figure 3.10*).

3.4.2 Deletions

Deletions of part of a gene are not an uncommon cause of genetic disease. To illustrate the methods used in the laboratory for determining the presence or absence of small or large deletions, we will respectively consider the most common mutation that causes cystic fibrosis in the UK, ΔF_{508}, and the deletions that can often cause Duchenne or Becker muscular dystrophy.

3.4.3 Deletions in cystic fibrosis

In England, about 75% of chromosomes in which the gene causing cystic fibrosis (cystic fibrosis transmembrane regulator, or CFTR) is mutated carry the same mutation. This is a deletion of three bases that code for phenylalanine, and the mutation is known as ΔF_{508} since the phenylalanine is at position 508 in the CF gene. This makes initial analysis of families with cystic fibrosis easy, as they can be investigated for the presence of ΔF_{508}. The simplest way of doing this is to amplify a fragment around the mutation by PCR, usually about 50 bp without the deletion. The PCR products can then be run on a polyacrylamide gel (see Section 2.4.2), which is stained with EtBr or silver stain. Chromosomes carrying the deletion will give a 47 bp fragment, while those without the deletion will produce a 50 bp band.

Although this initial step is easy for CF, if one or both chromosomes do not carry the deletion, the situation is much more difficult. The next five most common CF mutations account for about 5% of mutations, but that means 20% will not be accounted for. Laboratories specializing in CF mutation analysis will be able to detect more mutations, but there are now over 400 mutations described for CF, many of them only found in single families. Linkage analysis (Section 3.2.1) is therefore used quite often in CF families to identify individuals at risk, even though the gene is so well characterized. For a description of how 31 mutations in the CF gene can be detected in one reaction, see Section 3.8.

3.4.4 Deletions in Duchenne and Becker muscular dystrophy

Duchenne muscular dystrophy (DMD) is a very severe muscle wasting disease that usually results in death in the early twenties. Becker muscular dystrophy is a milder form of the disease caused by mutations in the same gene. It is X-linked, so the mutation is carried by females but manifests serious symptoms only in males. Around 65% of cases are due to deletions in the dystrophin gene, which is on chromosome Xp21. The other 35% of cases are due to other mutations, such as point mutations, in the same gene. The DMD gene is the largest human gene isolated, at over 2 million bases. There are three ways of finding deletions in the gene. One is to use multiplex PCR (see Section 3.4.1 under ARMS). This is the process whereby several primer pairs for different parts of the gene are used in the same tube. By analyzing all the places in the DMD gene where deletions have been found to occur, it has been possible to design 11 primer pairs that will detect 99% of deletions. This is fine for males who need to be investigated, but there is a problem with female carriers, where the nonmutated gene on the other X will mask the deletion by producing PCR fragments in the reaction. Some laboratories are able to discern the loss of signal, but this is not easy. To overcome this problem, two methods are available; PFGE (see Section 2.4.1), and fluorescent in situ hybridization, FISH (see Section 2.5.2). PFGE enables very large DNA fragments to be separated on agarose gels, which are then Southern blotted and probed with a radioactive piece of DNA from the DMD gene (see Section 2.1.3). A deletion may produce a different sized fragment from that seen normally. Another way to see deletions in females, if these mutations are large enough, is to use FISH on chromosome metaphase spreads with a probe that is within the deleted area. If there is no deletion there will be a signal on both X chromosomes, if there is a deletion, only one X chromosome will show a fluorescent spot.

3.4.5 Diseases caused by triplet repeat mutations

There are several genetic disorders that are caused by an expansion in a number of base triplets. *Table 3.1* gives the names of some of these diseases, which repeat is implicated, and the number of repeats which indicates the presence of the disease in each case. In all these conditions the number of repeats varies in the population, and only becomes pathogenic over a certain threshold. How these repeats cause the diseases is not clear, but they are an unusual class of mutation in that the number

Table 3.1: Some diseases with trinucleotide expansions

Disease (and chromosome)	Type of repeat (and position in the gene)	Inheritance pattern	Number of repeats in disease (and normal)
Huntington's chorea (4p)	CAG (coding)	Autosomal dominant	37–121 (10–34)
Spinocerebellar ataxia 1 (6p)	CAG (coding)	Autosomal dominant	41–81 (6–39)
Machado-Joseph disease (14q)	CAG (coding)	Autosomal dominant	68–79 (13–36)
Dentorubralpallido-luysian atrophy (12p)	CAG (coding)	Autosomal dominant	49–88 (7–25)
Spinal and bulbar muscular atrophy(X)	CAG (coding)	X-linked recessive	40–62 (11–34)
Fragile X syndrome (X)	CGG (noncoding - 5′ UTR)	X-linked recessive	200–2000 (6–52)
Myotonic dystrophy (19q)	CTG (noncoding)	Autosomal dominant	100-several thousand (5–37)
Friedreich's ataxia (9)	GAA (noncoding - intron 1)	Autosomal recessive	200–900 (7–22)

(5′ UTR is the 5′ untranslated region of the gene.)

of repeats tends to increase from one generation to the next, and this produces an earlier onset of symptoms and increased severity. This is known as clinical anticipation.

Huntington's chorea
Table 3.1 shows that there is one class of disease where the mutations are all in the coding regions of the gene and are CAG repeats that code for the amino acid phenylalanine. All these diseases are neurodegenerative. Huntington's chorea will be used as an example to show how the DNA is analysed. There are between 11 and 34 CAG repeats in the Huntington gene in unaffected individuals. Those with the disease have 37 to 86 repeats. It is possible to PCR directly across the repeat region, run the products on acrylamide gels, and determine the size of the insert. There are a few cases with intermediate alleles of 35 or 36 repeats, and these represent a diagnostic problem. Rarely, the repeat can get smaller. Before the gene was discovered in 1993, linked markers were used. This meant that there was always a possible error, however small. Now that the mutation is examined directly this error no longer exists.

Some potential carriers of the Huntington gene do not want to know whether they are affected, but they want unaffected children. This is possible using a method called prenatal exclusion. This is illustrated in *Figure 3.11*. The affected male in generation I does not have his DNA examined in this method. It is no use looking at the CAG repeat size in the fetus as this would immediately give information about the status of the mother. Instead of analyzing the gene, linked markers are used. In *Figure 3.11*, the female in generation I is homozygous for the marker used, and is

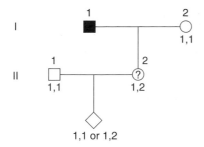

Figure 3.11: Pedigree to illustrate prenatal exclusion. Numbers above each person are for reference purposes. Numbers below each individual are markers close to the Huntington's chorea gene. The diamond shape represents a fetus of unknown sex. Individual II2 does not know her disease status and does not want to. If the foetus has the markers 1,1 it is unaffected, if it is 2,1 there is a 50% risk that it has the disease gene (with an error that is the recombination fraction for the marker used and the disease gene, see Section 3.2.2).

1,1. The mother of the fetus has alleles 1,2, and her partner has alleles 1,1. We therefore know that in the mother of the fetus allele 2 comes from her affected father, but we do not know whether this allele indicates the mutant or nonmutant gene. For prenatal exclusion this knowledge is not necessary. In *Figure 3.11*, the fetus can have two possible genotypes for the linked marker; 1,1 or 1,2. In the former case there is no allele from the affected grandfather, therefore the fetus is unaffected assuming there is no recombination between gene and marker (see Section 3.2.2). For Huntington's chorea this error is very small as there are many markers close to the gene. If the fetus has genotype 1,2 then there is a contribution from the grandfather, but it is not known whether this is the mutated gene or the nonmutated gene, the odds are equal. Thus, 50% of fetuses terminated do not have Huntington's chorea.

Fragile X and myotonic dystrophy
Both these diseases are caused by large triplet repeat expansions in the noncoding region of the gene. For fragile X it is a CGG repeat at the beginning of the gene, for myotonic dystrophy a CTG repeat at the end of the gene. Up to 100 repeats can be analyzed by PCR, as for Huntington's chorea, but for larger expansions Southern blotting has to be used to detect increased band sizes.

3.4.6 Duplications – Hereditary motor and sensory neuropathy 1A (Charcot Marie Tooth syndrome)

This autosomal dominant disease is often associated with a 1.5 Mb duplication of part of chromosome 17p11.2, and affects the muscles of both limbs. It is believed that doubling the copy of the peripheral myelin protein gene (PMP22), which lies within the duplicated region, is responsible for the disorder. The extra copy of the gene can be detected by southern blotting using a DNA probe from the PMP22 gene. Alternatively, PFGE can be used to detect the entire 1.5 Mb duplication.

3.4.7 Imprinting – Angelman's syndrome and Prader Willi syndrome

These two conditions are mostly sporadic, and 70% of each are associated with deletions on 15q11-13. Both diseases are neurobehavioral disorders, with Angelman's (AS) believed to be caused by mutations in the *UBE3A* gene, while the *necdin* gene is a candidate for Prader Willi (PWS). These syndromes are interesting from a diagnostic viewpoint because they are both examples of imprinting disorders. Imprinting, in this instance, is when the copy of the gene from either the mother or the father is inactive. The mechanism for this is unclear, but there appears to be a correlation with methylation of the bases in DNA, where the methylated gene is not transcribed into RNA. In PWS the maternal gene is methylated, and therefore inactive, so that deletion of the paternal gene causes the disorder. For AS, the imprinting is reversed and deletion of the maternal gene gives rise to the symptoms. There is a rapid diagnostic test based on parent-of-origin specific methylation differences. There is a *Not1* restriction site at the small nuclear riboprotein (SNRPN) locus, which is close to both AS and PWS loci. The *SNRPN* gene is methylated on the maternally derived chromosome, but unmethylated on the paternally derived chromosome. DNA is digested with restriction enzymes *Xba1* and *Not1*. The latter enzyme only cuts if the DNA is unmethylated. Southern blotting and probing with a fragment of the *SNRPN* gene produces a 4.2 kb band and 0.9 kb paternal band. In PWS there is only a 4.2 kb band (paternal unmethylated is missing), and in AS there is only a 0.9 kb band (maternal methylated is missing). This methylation method has been adapted to a PCR format.

3.5 Karyotyping for trisomies and translocations

The introduction of FISH (see Section 2.5.2) has made the detection of chromosomal rearrangements much easier to see. Translocations, where part of one chromosome breaks and becomes attached to another, were sometimes difficult to see using the older staining systems, but even small rearrangements are visible with FISH. The most common chromosomal abnormalities are those of number, such as Down's syndrome, which is due to the presence of an extra copy of chromosome 21. This can be seen easily on a metaphase spread by a cytogeneticist. The most frequent translocations are the so called centric fusion, or Robertsonian, variety. These occur when there are breaks near the centromere in two acrocentric chromosomes , with cross fusion of the products. Acrocentric chromosomes are those with centromeres near one end, 13, 14, 21, and 22 (see *Figure 2.3*). The Robertsonian translocation seen most often is that between chromosomes 13 and 14. These can be detected by G banding or chromosome painting (see Section 2.5.2). Although a small amount of DNA is lost from each chromosome, there is no effect on the individual. Owing to miss-pairing there can be problems at gametogenesis, however.

3.6 Single cell detection of genetic disorders in embryos by PCR

There are occasions when pre-implantation diagnosis of embryos is called for, for example, when mothers are infertile and have to use *in vitro* fertilization (IVF), and there is a genetic disease in the family, or where a mother has had repeated miscarriages due to a chromosomal translocation. This involves taking just one or two cells from the embryo and analyzing them, either by PCR or FISH. Single cell PCR presents huge problems of possible contamination, but has been shown to be possible if great care is exercised. FISH avoids any problems of contamination, but gives at best only two cells, and quite often only one to work with, with no amplification step as in PCR. Unlike the situation when cells are cultured for prenatal diagnosis and can be stopped in metaphase, it is not possible to analyze condensed chromosomes in FISH. The analysis has to be done in interphase. To detect whether a translocation is balanced or unbalanced in interphase, probes that bind to either side of the breakpoint on one chromosome are used, with different colors for each probe, in conjunction with a third probe that maps anywhere on the other chromosome. If there are two signals for each probe then the translocation is balanced, but any other combination shows that it is unbalanced. This means that for each case probes have to be sought that map either side of the breakpoints, but as they can be at any distance they can sometimes be used for more than one family.

A translocation is balanced if the correct amount of genetic material is present, and unbalanced if there is duplication and/or deletion of part of one or more chromosomes.

3.7 Polygenic disorders

The most common genetic disorders are not those caused by single gene mutations, but are due to the influences of many genes plus environmental factors. Such diseases are called polygenic or multifactorial, and include hypertension and coronary artery disease. These are the disorders that in the future will be understood well enough to be analyzed using nucleic acid diagnostic methods, but at the moment this is not the case. It is known that certain genes are involved in these disorders, but the work is still at the research stage.

3.8 Automated analysis for common mutations

Automated DNA sequencers are now a common piece of equipment in DNA diagnostic laboratories. These use lasers to detect different fluorescent dyes. Four different dyes are currently available, which means that many samples, such as CA repeats or different mutations, can be run in one lane of a gel and analyzed by the sequencer. As long as those lengths of DNA of a similar size have a different dye the machine can distinguish between them. For example, there is a kit available that allows analysis of the 31 most common mutations in the *CFTR* gene (that cause cystic fibrosis) in one lane of the sequencer. This is achieved by designing primers for the different mutations that have a range of product sizes, and any that are of a similar size have a different dye label. As each gel has 36 lanes, these sequencers are invaluable for screening large numbers of samples or many mutations in individuals.

3.9 Associations of particular alleles and disease states in the population

Sometimes there is an association between polymorphisms in certain genes and genetic disease. Two very good examples of this are HLA antigens and insulin dependent diabetes, and ApoE and Alzheimer's disease. HLA stands for human leukocyte antigens, which are membrane proteins involved in the presentation of antigens to immune cells (an antigen being anything foreign to the body, such as viral or bacterial proteins). The *HLA* genes are clustered together on chromosome 6, and show different forms in the population. About 95% of patients with type 1 diabetes in the UK have HLA DR3 or DR4, compared to 50% in the population. Even more striking is the association of HLA B27 with the disease of ankylosing spondylitis. In whites in the UK this is found in 95% of patients but only 7% of the population. Not all individuals with this HLA type will develop the disease, but the risk is 90 times that of those without HLA B27. There is a gene called apolipoprotein E which has several different forms. One of these, called *apoE* epsilon, shows a striking association with late onset Alzheimer's disease. The exact nature of these associations is not clear but by analyzing the various genes it is possible to produce a risk assessment for different individuals. This can be a relatively small increase in risk, as in the case for HLA and diabetes, or a substantial increase, as in ankylosing spondylitis. The ethical questions involved in this type of analysis are considerable, as there is no certainty of an individual developing the disease. As with all developments in genetic analysis, they need to be used for the benefit of the population and not as a means of screening to divide into different genetic classes of individual.

3.10 Cancers

Detection of nucleic acid changes in histopathological specimens will be described in Chapter 6. A brief description is given below of some of the familial forms of cancer, and the diagnostic techniques used in these conditions.

3.10.1 Familial breast cancer genes, BRCA1 and BRCA2

A high percentage of familial breast and ovarian cancers are caused by mutations in the *BRCA1* or *BRCA2* genes. The disease is inherited in an autosomal dominant manner. These are large genes with 22 and 26 coding exons, respectively. Mutation analysis can involve the PTT (see Section 2.5.4), SSCP (see Section 2.4.3), direct sequencing, or ARMS. (see Sections 2.4.5 and 3.4.1). For PTT either RNA or DNA can be used. The technique is simpler with DNA as a template, but there needs to be one or more large exons in the gene for this to be feasible. For BRCA1 this is the case, as exon 11 codes for over half the gene. This method will only detect stop codons, and if it does not show up a mutation the exons have to examined with SSCP. For populations where particular mutations are known to be more common, such as the 185 delAG mutation in exon 2 (where nucleotides A and G are deleted at position 185) in Ashkenazi Jews, this will be the first line of analysis. Either ARMS primers specific for this mutation or direct sequencing of the relevant exon can be used.

3.10.2 Familial adenomatous polyposis coli (FAP)

FAP is a rare form of colon cancer, again autosomal dominant, where there are mutations in the *APC* gene. About 95% of mutations in FAPC are premature stop codons, and of these 65% cluster in exon 15, although this only accounts for about 10% of the coding region. PTT for this exon is therefore the first method used (see Section 2.5.4), followed by SSCP (see Section 2.4.3) and sequencing of other exons if the PTT is negative (see Section 2.4.5).

3.10.3 Hereditary nonpolyposis colorectal cancer (HNPCC)

In this autosomal dominant form of bowel cancer there are no preceding multiple polyps as in FAPC. Mutations in MSH2 and MLH1, genes that code for mismatch repair proteins, are responsible for about 90% of cases. At risk individuals have one normal copy of this gene, which is perfectly adequate, but tumors arise when somatic mutations occur in this gene, leaving the cell with no functioning repair gene. PTT from RNA is often used to look for any stop mutations in this gene (see Section 2.5.4), followed as above by SSCP (see Section 2.4.3) of the coding exons if there is no mutation shown by PTT.

3.10.4 Cowden's disease

A rare autosomal dominant cancer predisposition syndrome, with an increased risk of breast, thyroid and skin tumors. The main clinical symptom is benign hamartomas of multiple organs. Mutations in the tumor suppressor gene *PTEN* are responsible for the disorder. There are only nine exons in this gene, so SCCP or DGGE (see Section 2.4.3) are the usual detection methods.

3.10.5 Multiple endocrine neoplasia (MEN) and familial medullary thyroid carcinoma (FMTC)

MEN 2A, MEN 2B and FMTC are all inherited cancer syndromes produced by mutations in the same gene. MEN 2A is characterized by medullary thyroid carcinomas, hyperplasia of the parathyroid and pheochromocytoma (an adrenal tumor). MEN 2B is similar, but includes mucosal neuromas and lacks parathyroid hyperplasia, while FMTC has just medullary thyroid tumors. All three conditions are caused by gain of function mutations in the RET (a receptor tyrosine kinase) gene. There are specific mutations associated with each condition, but mutations elsewhere in the gene can be responsible.

MEN1 is distinguished by the presence of pancreatic islet cell tumors, and the gene for this form of the disease, known as the *MEN1* gene, is a tumor suppressor gene on chromosome 11q13. There are 10 exons in this gene, so that SSCP (see Section 2.4.3) is the most common method used for mutation detection.

3.10.6 Familial melanoma

The p16 gene that is mutated in some cases of familial melanoma is a cyclin dependant kinase inhibitor. The product of this gene acts in a complex pathway that regulates cell division, and when mutated a 'brake' on the system is removed. This can lead to uncontrolled cell growth, and thus produce melanomas. The p16 gene has only seven coding exons, so it is possible to use SSCP (see Section 2.4.3) or even direct sequencing of all the exons (see Section 2.4.5) to look for mutations.

Further reading

Beggs, A.H., Koenig, M., Boyce, F.M. and Kunkel, L.M. (1990) Detection of 98% of DMD/BMD gene deletions by polymerase chain reaction. *Hum. Genet.* **86:** 45–48.

Birren, B. and Lai, E. (1990) *Pulsed field gel electrophoresis: A practical guide.* Academic Press. San Diego.

Conn, C.M., Cozzi, J., Harper, J.C., Winston, R.M. and Delhanty, J.D. (1999) Preimplantation genetic diagnosis for couples at high risk of Down syndrome pregnancy owing to parental translocation or mosaicism. *J Med Genet* **36:** 45–50.

Davies, K. (ed) (1993) *Human genetic disease analysis: A practical approach.* Information Press. Oxford.

Delach, J.A., Rosengren, S.S., Kaplan, L., Greenstein, R.M., Cassidy, S.B. and Benn, P.A. (1994) Comparison of high resolution chromosome banding and fluorescence in situ hybridisation (FISH) for the laboratory evaluation of Prader-Willi syndrome and Angelman syndrome. *Am. J. Med. Genet.* **52:** 85–91.

Ferrie, R.M., Schwarz, M.J., Robertson, N.H., Vaudin, S., Super, M., Malone., G. and Little, S. (1992) Development, multiplexing and application of ARMS tests for common mutations in the CFTR gene. *Am. J. Hum. Genet.* **51:** 251–262.

Froggatt, N.J., Green, J., Brassett, C., Evans, D.G., Bishop, D.T., Kolodner, R. and Maher, E.R. (1999) A common MSH2 mutation in English and North American HNPCC families: origin, phenotypic expression, and sex specific differences incolorectal cancer. *J. Med. Genet. 36:* 97–102.

Ginot, F. (1997) Olgonucleotide micro-arrays for identification of unknown mutations: how far from reality. *Hum. Mut.* **10:** 1–10.

Gortz, B., Roth, J., Krahenmann, A., de Krijger, R.R., Muletta-Feurer, S., Rutimann, K., Saremaslani, P., Speel, E.J., Heitz, P.U. and Komminoth, P. (1999) Mutations and allelic deletions of the MEN1 gene are associated with a subset of sporadic endocrine pancreatic and neuroendocrine tumors and not restricted to foregut neoplasms. *Am J Pathol* **154:** 429–36.

Julier, C., Hashimoto, L. and Lathrop, G.M. (1996) Genetics of insulin dependant diabetes mellitus. *Curr. Opin. Genet Dev.* **6:** 354–360.

Liaw, D., Marsh, D.J., Li, J., Dahia, P.L.M., Wang, S.I., Zheng, Z., Bose, S., Call, K.M., Tsou, H.C., Peacocke, M., Eng, C. and, Parsons, R. (1997) Germline mutations of the PTEN gene in Cowden's disease, an inherited breast and thyroid cancer syndrome. *Nat. Genet.* **16:** 64–67.

MacKie, R.M., Andrew, N., Lanyon, W.G. and Connor, J.M. (1998) CDKN2A germline mutations in U.K. patients with familial melanoma and multiple primary melanomas. *J. Invest. Dermatol.* **111:** 269–72.

Roses, A.D. (1998) Apolipoprotein E and Alzheimer's disease. The tip of the susceptibility iceberg. *N.Y. Acad. Sci.* **30:** 738–743.

Ross, C.A., McInnis, M.G., Margolis, R.L. and Li, S.H. (1993) Genes with triplet repearts:candidate mediators of neuropsychiatric disorders. *TINS* **16:** 254–260.

Serova, O.M., Mazoyer, S., Puget, N., Dubois, V., Tonin, P., Shugart, Y.Y., Goldgar, D., Narod, S.A., Lynch, H.T. and Lenoir, G.M. (1997) Mutations in BRCA1 and BRCA2 breast cancer families: are there more breast cancer-susceptibility genes? *Am. J. Hum. Genet.* **60:** 486–495.

Shirahama, S., Ogura, K., Takami, H., Ito, K., Tohsen, T., Miyauchi, A. and Nakamura, Y. (1998) Mutational analysis of the RET proto-oncogene in 71 Japanese patients with medullary thyroid carcinoma. *J. Hum. Genet.* **43:** 101–6.

Strachan, T. and Read, A.P. (1996) *Human Molecular Genetics.* Bios scientific publishers.

Trent, R.J. (1997) *Molecular Medicine.* 2nd Edn. Churchill Livingstone Publishers.

Van der Luit, R., Khan, M., Vasen, H., van Leeuwen, C., Tops, C., Roest, P., den Dunnen, J. and Fodde, R. (1994) Rapid detection of translation-terminating mutations at the adenomatous polyposis coli (APC) gene by direct protein truncation test. *Genomics* **20:** 1–4.

Weber, J.L. and May P.E. (1989) Abundant class of DNA polymorphisms which can be typed using the polymerase chain reaction. *Am. J. Hum. Genet.* **44:** 388–396.

Wetherall, D.J. (1991) *The new genetics and clinical practice.* 3rd Edn. Oxford Medical Publications.

Willems P.J. (1994) Dynamic mutations hit double figures. *Nat. Genet.* **8:** 213–215.

Zeschnigk, M., Lich, C., Buiting, K., Doerfler, W. and Horsthemke, B. (1997) A single tube PCR test for the diagnosis of Angelman and Prader-Willi syndrome based on allelic methylation differences at the SNRPN locus. *Eur. J. Hum. Genet.* **5:** 94–98.

Chapter 4

Infectious diseases I – viruses

4.1 Introduction

Molecular probes are gaining wide acceptance as the techniques of choice for the detection of many micro-organisms. They have the advantages over conventional methods of speed, specificity and sensitivity. In addition they may be the only applicable techniques as many micro-organisms are fastidious and cannot be easily grown for phenotypic analysis in the laboratory. This is particularly pertinent to the study of viruses which, as obligate intracellular parasites, require mammalian cells for their culture. Even when the appropriate cell type is used, it is often not possible to achieve the same *milieu* that obtains in the natural environment for efficient growth. Apart from detection, molecular methods are now becoming standard technologies for establishing viral load, particularly with regard to monitoring treatment, and resistance to antiviral therapy. *Table 4.1* shows a comparison of molecular probes with more conventional methods.

Table 4.1: A comparison of molecular probes with conventional methods for the detection of micro-organisms

	Culture	Immuno-fluorescence	ELISA	Non-amplification probes	Gene amplification methods
Speed to produce result	+	+++	+++	++	++/+++
Sensitivity	+++	++	++	++	++++
Specificity	+++	++	++	+++	++++
Quantifiability	++	++	++	+	+++
Ease of use	+	+	+++	+	++/+++

Viruses differ significantly from other infectious agents. They are obligate intracellular parasites which exist extracellularly as chemicals. They possess either RNA or DNA unlike all other infectious agents which possess both. From the viewpoint of probe technologies it is useful to know the type of target nucleic acid (*Table 4.2*). In theory it should be

57

Table 4.2: Types of genome possessed by specific viruses

DNA	Parvoviruses, e.g. B19
	Hepadnaviruses, e.g. hepatitis B virus
	Papovaviruses, e.g. papillomavirus
	Adenoviruses Herpesviruses, e.g. cytomegalovirus
	Poxviruses, e.g. orf virus
RNA	Picornaviruses, e.g. rhinovirus
	Coronaviruses, e.g. human coronavirus 229E
	Orthomyxoviruses, e.g. influenza
	Paramyxoviruses, e.g. Respiratory Syncytial virus
	Reoviruses, e.g. rotavirus
	Flaviviruses, e.g. hepatitis C
	Caliciviruses, e.g. Norwalk virus
	Rhabdoviruses, e.g. rabies virus
	Togaviruses, e.g. rubella viruses
	Arenaviruses, e.g. Lassa virus
	Filoviruses, e.g. Ebola virus
	Bunyaviruses, e.g. Sandfly fever virus
	Retroviruses, e.g. HIV

possible to diagnose virus infections rapidly by electron microscopy and culture. In practice, the former is insensitive, requiring about 10^6 virions to be present in a sample, and the latter slow and not always possible. Molecular probes, and gene amplification, overcome these problems and have become routine procedures in the diagnostic setting.

4.2 Sample collection and preparation

Nucleic acid from DNA viruses can be processed in a manner not dissimilar to that used for DNA from eukaryotic and bacterial sources. RNA is, however, more labile and specimens containing viral RNA need to be handled rigorously. Containers should be both sterile and treated with diethylpyrocarbonate to minimize RNAse activity. Collection should be undertaken with gloves as RNAse activity is high in sweat and other bodily fluids. The use of RNAse inhibitors may also be required if viral load is low or transit conditions to the laboratory are less than ideal. Extraction of RNA as template requires rigorously controlled conditions and most often employs a guanidium-isothiocyanate (GI) extraction step [1]. Commercial methods based on GI extraction are now available.

4.3 Detection of virus

Almost all human viruses have had probe methods described for their detection. The whole gamut of test formats have been used: DNA:DNA hybridization, RNA:DNA hybridization, RNA:RNA hybridization, dot-blot, slot-blot and gene amplification. Generally, however, amplification methods are now employed.

4.3.1 PCR methods

PCR has had widespread applicability to the detection of viruses in range of specimen types. For DNA viruses this is straightforward, for RNA viruses an additional reverse transcription step is required either by a separate RT enzyme or a DNA polymerase, such as *Thermus thermophilus* DNA polymerase, which has RT activity. Under appropriate conditions the use of the single enzyme is not only more convenient but has been shown to be more efficient, in the case of amplification of hepatitis C virus from human plasma a hundred times as efficient [2].

For some viruses nucleic acid probes are far superior to the alternatives. Human papillomaviruses (HPV) cannot be routinely grown in cell culture and serological assays, now being developed, are beset by poor sensitivity and cross-reactivity of antigens. These viruses are recognized by their genotype of which over 60 are now recognized. It is impractical to perform a large number of Southern or Northern blots and so dot-blots formats have been developed. These are sensitive and can detect as little as 10 pg DNA. The reason for wanting to detect and type papillomaviruses is to determine cancer risk. Certain types of HPV, such as 16 and 18, are strongly associated with the risk of cervical cancer. Other types may be more common but are not oncogenic. Commercial systems are now available that detect all the common types and have a near 100% sensitivity for detecting those that are oncogenic [3]. These viruses were first identified as the cause of warts but the use of gene probes has identified that clinically apparent infections (which may still be transmitted and be oncogenic) are the more common manifestation. PCR studies have shown that up to half of Papanicolou smear-negative cervical specimens carry HPV, including type 16, suggesting that this is a more sensitive test for determining the risk of subsequent cervical cancer [4]. PCR also allows genotyping typing with the use of type-specific primers. Other viruses that cannot be grown routinely are shown in *Table 4.3*. Nucleic acid probes are not the initial method of choice for all these viruses, however, as antigen detection, for example for hepatitis B virus, is sufficiently sensitive at the clinically relevant time in most cases. Even with hepatitis B, however, PCR has a role as the detection of viral DNA in the plasma of carriers who have equivocal HBeAg/HBeAg as a positive result is a marker of high infectivity. Additionally viral variants with mutations in the core gene may be negative for HBeAg because there is a premature stop codon in the open reading frame: PCR can be used to show that these patients are highly infectious and are falsely HBeAg-negative.

Some viruses can be routinely grown in cell culture but grow so slowly as to be clinically unhelpful. An example of this is the cytomegalovirus (CMV). An innocuous virus mostly, it can cause retinitis and life-threatening pneumonia in the immunocompromised. Standard cell culture techniques can take up to 3 weeks for positive isolation although this can

Table 4.3: Examples of common viruses that cannot be routinely grown in cell culture

Virus	Disease
Hepatitis A	Infectious hepatitis
Hepatitis B	Infectious hepatitis
Hepatitis C	Infectious hepatitis
Parvovirus B19	Erythema infectiosum
Coronaviruses	Common cold
Enteroviruses	Meningitis, rashes and other infections
Polyomaviruses	Progressive multifocal leukoencephalopathy
Papillomaviruses	Warts
Epstein-Barr virus	Infectious mononucleosis

be speeded up by immunofluorescent or enzymatic detection of early antigens in 24–48 hours. Both simple probe and amplification methods have been applied to the detection of CMV. The extreme sensitivity of PCR is useful if the clinical features are suggestive but if these are ambiguous may cause confusion. Positive results may occur because of latent virus which is common in adults. The likeliest benefit will come from examining material from neonates and clinical specimens that would not normally harbor the virus. In one study in which primers were used that hybridized to regions of the MIE gene (the 'major immediate early' antigen gene, an expressed nonstructural product), all 44 culture-positive neonates were PCR positive in their urine but none of the 27 culture-negative neonates were [5]. The most problematic specimen type is blood as the virus is lymphotropic and latent virus can be detected in white cells. Unless there is bleeding into the meninges, however, CSF would not normally be expected to harbor latent virus and its use on CSF in AIDS patients with neurological disorders has shown PCR to be able to detect almost all cases with disease even though cell culture was more commonly negative [6].

With viruses expected to be present in low concentrations, nested PCR is often used and adds to specificity; the increased sensitivity also, however, makes the risk of false positives due to contamination greater. With specimens which may have a number of different viruses or even mixed infections, multiplex PCR methods have been developed. Clearly the design of multiplex PCR needs to be meticulous with the melting temperature (T_m) of primers being similar and the products being of differing size so as to be distinguishable. Specimens such as CSF and respiratory secretions are ideal samples for the use of multiplex PCR as the clinical features are unlikely to identify the causative virus. Another use of multiplex PCR is to type viruses. There are, for example, four serotypes of dengue virus which can cause an illness ranging from simple fever to hemorrhagic fever. It is useful to be able to type this virus as it is recognized that infection with one serotype, following prior infection with another, is a risk factor for the serious hemorrhagic fever manifestation.

A single tube multiplex RT-PCR method has been shown to be able to detect and type 1–$50\,\text{pfu ml}^{-1}$ of virus in clinical specimens and the mosquito vector [7]. (NB. A pfu is a plaque forming unit, and is used as a measure of the amount of viable virus present in culture. One plaque represents one viable viral particle.).

4.3.2 Non-PCR methods

Although PCR methods have been the most commonly used, other nucleic acid amplification technologies have also been applied to viruses. In particular, human immunodeficiency virus (HIV) has been a popular target. It is a dangerous pathogen and, even if culture was not technically demanding, safety is a good reason for using molecular probes. Target amplification systems such as self-sustained sequence replication (3SR) have been used to amplify a 214 bp region of the *env* gene of zidovudine-resistant HIV[8]. Probe amplification techniques such as ligase chain reaction (LCR) (*Figure 4.1*) and Qβ replicase (*Figure 4.2*) have also been used and are being developed commercially. LCR is useful for the detection of mutations which are frequent in HIV [9]. It depends on the ability of a thermostable DNA ligase to seal nicked double stranded DNA. If two primer probes bind contiguously to the target, this situation occurs. Repeated hybridization/disassociation cycles result in multiple copies of the probe pair. Qβ replicase is an enzyme derived from the Qβ RNA bacteriophage. This enzyme is capable of replicating a 221 bp RNA named MDV-1. If MDV-1 containing a probe is bound to target nucleic acid, this results in multiple copies of the recombinant. Qβ replicase amplification has been shown to be able to detect low level HIV-1 infection although it appears to be less sensitive than PCR because of high background amplification of unhybridized probe recombinant [10]. Its advantages are that it can detect both RNA and DNA targets and is more easily quantifiable than PCR.

4.3.3 Molecular epidemiology

Viruses evolve through mutation and recombination events at a rate much faster than living organisms; this is greater in those with an RNA genome. It is helpful to be cognizant of this variation for a number of reasons: to identify the spread of outbreaks; to understand the natural epidemiology of viruses including animal reservoirs; and to develop more effective vaccines. As genetic variation is a precursor of antigenic variation, subtler differences can best be detected by genome analysis. Multisegmented viruses such as rotaviruses exhibit gross changes by variation in size of those segments: these can be visualized by agarose or polyacrylamide gel electrophoresis. The genomes of large DNA viruses can be analyzed by RFLP mapping if sufficient virus is available. Oligonucleotide

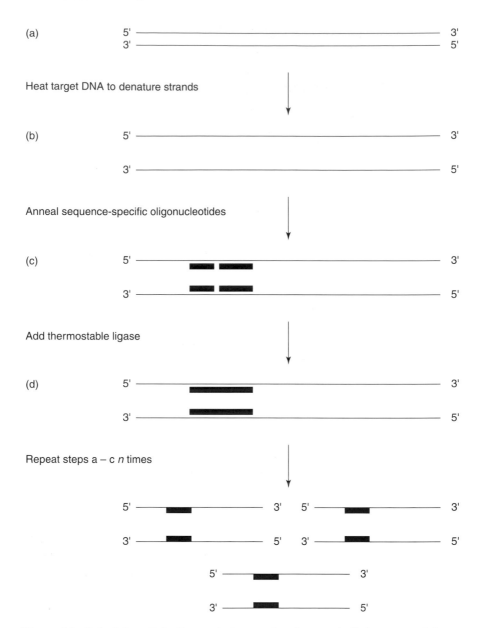

Figure 4.1: Principles of the ligase chain reaction (see text). Only one straight detection shown.

fingerprinting is a useful method for typing RNA genomes. Digestion of the genome with an enzyme, such as ribonuclease T1, for defined periods results in fragments which can be separated by two-dimensional electrophoresis. If radiolabeled the fragments can be visualised by autoradiography to form a characteristic pattern.

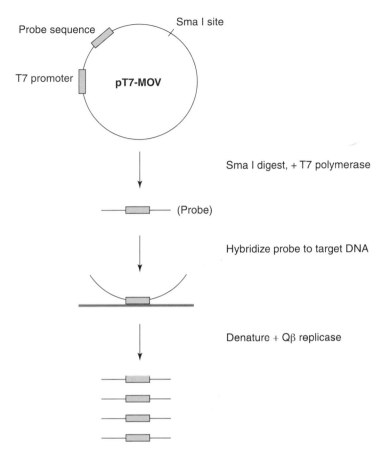

Figure 4.2: Principles of Qβ replicase probe amplification system (see text).

DNA probes are useful for determining defined sequence variants but as with diagnostics, PCR has offered a more rapid and sensitive approach. Two methods have been commonly employed, PCR-restriction endonuclease (PCR-RE) and PCR-single-strand-conformation polymorphism (PCR-SSCP) analysis. The former is also termed PCR-restriction fragment length polymorphism (PCR-RFLP) analysis and is restriction enzyme digestion of PCR amplicons. Its use has been described to type strains of hepatitis B [11]. PCR-SSCP employs a nondenaturing polyacrylamide gel to compare melted single strands of amplicons. In theory a strand with a single base mutation will migrate differently from the nonmutant type. In practice this degree of resolution requires meticulous technique

Methods such as RAPD used for bacterial typing could also be applied to larger viral genomes and gene sequencing applied for the ultimate discrimination (see Chapter 5).

4.4 Quantitative viral estimation

Quantification of viral load is becoming increasingly important as a means of monitoring antiviral therapy of viruses such as HIV and hepatitis C. It is also used, in HIV infection, as a means for deciding when to start therapy. PCR, nucleic acid sequence-based amplification (NASBA) and branched chain DNA (bDNA) amplification methods have all been applied to quantification of viral load in the clinical setting: LCR and Qβ could also be applied.

Quantitative competitive PCR includes a target sequence mimic (imitator) which contains a template (control sequence) which is amplified as efficiently as the actual target. Thus a comparison of, for example, HIV-1 target amplification with that of the control allows for a value to be extrapolated. There are a number of different methods that allow differentiation between the two sets of amplicons. The insertion of a restriction site, the inclusion of internal deletions or insertions, or replacement of a portion of the sought sequence by a novel sequence in the control are all used. The last is used in the commercially available system from Roche. The standard Roche assay for HIV-1 has a lower limit of detection of 400 copies ml^{-1} and a linear dynamic range of $10^{2.6}$–$10^{5.9}$ copies ml^{-1}. This can be enhanced by the use of 'real-time' PCR to 50 copies ml^{-1}.

NASBA is an isothermal RNA amplification system which has a similar lower limit of detection to PCR. It is available commercially from Organon Teknika with a linear dynamic range of $10^{2.3}$–$10^{7.2}$ copies ml^{-1}. when applied to HIV-1 quantification. This is a transcription-based amplification system (TAS), as is the similar self-sustained sequence amplification (3SR), which utilizes 3 enzyme activities: RT, RNAse H and T7 RNA polymerase. An oligonucleotide probe primer is bound to target RNA and the RT makes a DNA copy. RNAase H removes the RNA portion of the RNA-DNA hybrid and allows a second probe primer to anneal downstream. RT then acts as a DNA-dependent DNA polymerase to extend from one probe binding site to the other. One probe primer has a T7 promoter site incorporated so that this enzyme can then produce a further RNA copy to allow the process to start again. Typically a 10^9 amplification can be achieved. Quantification is usually by competitive amplification with a control as with PCR. NASBA has an inherent advantage over PCR of not being as sensitive to differing amplification conditions and can be used with accurate quantification on a wide range of specimen types.

bDNA amplification does not require an internal control template to be quantifiable. It is a signal amplification method that uses branched chain DNA probes that can then act as substrates for further hybridization reactions if the template is present initially. The technology is licensed by Chiron and has a linear range for HIV-1 of 10^4–$10^{6.2}$ copies ml^{-1}. The choice of test format is clinically not important as trends are sought. Consistency of test format is, however, important as the viral load value is not equivalent for each format. The more sensitive assays will have an advantage if complete reduction of viral load ever becomes a possibility on therapy.

4.5 Measurement of antiviral resistance

Even conventional phenotypic assays based on cell culture have not until recently been used routinely. This has partly been because of the paucity of effective antivirals but also because antiviral resistance has not been frequently documented. This picture is slowly changing with more antivirals and increased use so that antiviral susceptibility testing is now available in reference centres. Genotypic determination of viral (as opposed to clinical) resistance relies on identification of mutations that confer this state. Thus, it is known that mutations in the UL97 phosphotransferase gene and the UL54 DNA polymerase genes of cytomegalovirus confer resistance to the antiviral drugs, ganciclovir and/ or foscarnet. For some viruses, such as HIV which mutates frequently like most other RNA viruses, a battery of probes could be used to look for the common mutations that confer resistance in the RT and protease genes. Use of PCR to detect a specific mutation in the RT gene of HIV-1 showed, for example, that the development of these mutations was linked to a reduced sensitivity to zidovudine [12].

4.6 Detection of novel agents of disease

Many human diseases are of unknown etiology. Although multifactorial, infection is suggested by the epidemiology as a factor. The construction of cDNA and genomic DNA libraries from infected material which can then be screened using probe sequences of known agents has become established as a method for identifying known viruses as potential pathogens of a specific disease manifestation. Alternatively expression libraries have been screened with antisera raised against purified proteins, or with sera from infected patients. When an unknown agent is suspected then subtraction hybridization, comparing DNA sequences from infected with noninfected material, has been used. A most dramatic example of this was the identification of hepatitis C virus, the commonest cause of non-A non-B hepatitis [13]. These techniques have required up to microgram quantities of extracted nucleic acid to generate adequate libraries of 10^5–10^6 clones. The development of amplification methods has made this process more efficient.

Nonspecific amplification (NSA), also termed sequence-independent single primer amplification (SISPA) has been successfully used to isolate an immunoreactive region of Norwalk virus genome from a λgt11 cDNA library, screened with sera from a patient convalescing from infection [14]. The principles of this method are shown in *Figure 4.3*. An enhancement of

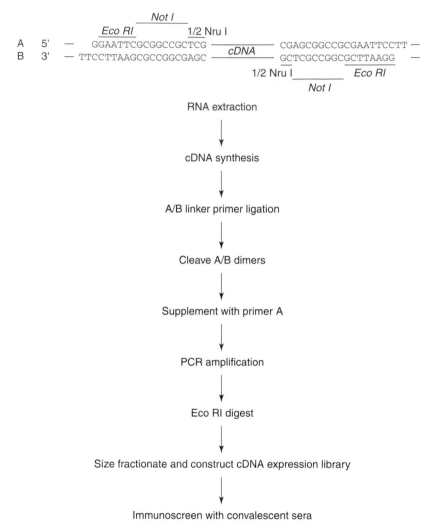

Figure 4.3: The principles of cDNA library construction by non-specific amplification (adapted from Reyes and Kim 1991).

this technique is random-PCR (rPCR). This is 100-fold more efficient and was used to identify a near-complete genome of a Japanese isolate of hepatitis C virus. rPCR is PCR using an oligonucleotide primer possessing a 20 nucleotide 'universal' sequence, which contains restriction enzyme recognition sites for cloning, and a random hexamer at the 3′ end. It obviates the need for the linker/primer step of NSA. A novel virus (termed GB) from patients with hepatitis have been cloned by NSA to produce a cDNA expression library and immunoscreened using antiserum from infected tamarin monkeys. Immunoreactive domains were identified [15]. This virus is currently an orphan virus and its role in causing hepatitis and/or other diseases is to be elucidated.

The identification of unknown viral pathogens has been achieved by subtractive technologies, allied to amplification methods. The principle of subtraction involves the removal of common sequences in infected and noninfected material with complex sequences. When two samples of broadly similar sequences are compared, the control source is termed the 'driver' and the experimental (infected) source termed the 'tester'. In general, a large excess of driver dsDNA is allowed to hybridize to single-stranded tester nucleic acid. Any nonhybridized single-stranded tester nucleic acid can be separated as presumed unique sequences. Originally, separation of single-stranded nucleic acid was by hydroxyapatite chromatography but modifications include the use of biotinylated driver DNA and oligo-dT latex particle separation. Subsequent gene amplification can be achieved by adding unique linkers to all nucleic acids. One of these methods, representational differential analysis (RDA) has been used to identify a novel herpesvirus from a vascular tumor common in AIDS patients, Kaposi's sarcoma [16]. RDA simultaneously combines amplification and subtraction. After restriction enzyme digestion to reduce genome complexity, linkers are added to the products to enable subsequent PCR amplification. Smaller fragments are amplified most efficiently. Thus the resulting amplicons may represent only a proportion (typically 10%) of the DNA restriction fragments. By the judicious use of single restriction enzymes for digestion of the genome, it is possible to produce amplicons that represent the entire genome. This is done for both tester and driver DNA. Subsequent steps in the RDA process are shown in *Figure 4.4*. RDA is being applied to a number of diseases of potential viral etiology.

The applicability of RDA is not, however, universal as the two nucleic acid sources required for the tester and driver need to be highly matched for practicability. Other methods, such as genome mismatch scanning might offer an alternative approach. This identifies regions of identity between two sources of nucleic acid. This is a method which has had use in linkage mapping of human genetic disorders (see Chapter 3). Suppression subtraction hybridization (SSH) is another alternative to RDA which is designed to selectively amplify differentially expressed cDNA and simultaneously suppress nontargeted DNA amplification.

The identification of a virus, or any other micro-organism, from patients with a disease is, of course, not sufficient grounds to establish cause and effect. Although not conclusive, the finding of nucleic acid within a diseased cell, but not a normal cell, provides convincing evidence of aetiology. *In-situ* hybridization enables this. As with other probe methods, *in-situ* PCR is becoming the advancement of choice, and has been applied to human papillomaviruses to elucidate their role in the genesis of cervical cancer [17]. *In-situ* PCR can be direct or indirect. Indirect methods involve adding standard PCR reagents to the fixed cells and then post-amplification the products are detected by standard *in-situ* hybridization. Direct *in-situ* PCR involves the use of a labeled dNTP in

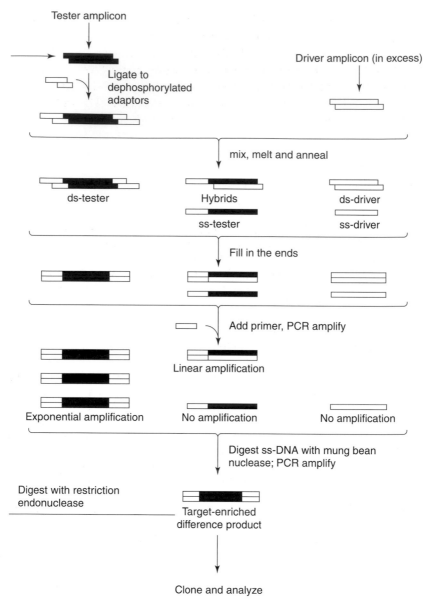

Figure 4.4: Principles of RDA.

the PCR reaction so that it can detected: a biotin- or digoxigenin dUTP detected by labeled avidin or antidigoxigenin are common examples.

4.7 Commercial systems

Commercial systems, such as Virapap®, Viratype® and HPV Profile® are available and used for the detection and typing of human papillomaviruses. These use a dot-blot hybridization format but most commercial developments have employed gene amplification technologies for speed and sensitivity. PCR and methods have been licensed by Roche for number of viral pathogens, including HIV-1 and hepatitis C. These utilize a biotinylated primer which can be captured onto a solid phase and then detected using a standard streptavidin-peroxidase ELISA system. Commercial quantification systems have already been described.

References

1. Chomczynski, P. and Sacchi N. (1987) Single-step method of RNA isolation by acid guanidium thiocyanate-phenol-chloroform extraction. *Anal. Biochem.* **162:** 156–159.
2. Young, K.K.R., Resnick, R.M. and Myers, T.W. (1993) Detection of hepatitis C virus RNA by a combined reverse transcription-polymerase chain reaction assay. *J. Clin. Microbiol.* **31:** 882–886.
3. Kiviat, N.B. *et al.* (1990) Comparison of Southern transfer hybridisation and dot filter hybridisation for detection of cervical human papillomavirus infection with types 6, 11, 16, 18, 31, 33 and 35. *Am. J. Clin. Pathol* 1990; **94:** 561–565.
4. Critchlow, C.W. and Koutsky, L.A. (1995) Epidemiology of human papillomavirus infection. In *Genital warts: human papillomavirus infection.* Mindel A (ed.), Edward Arnold, London, 53–81.
5. Demmler, G.J., Buffone, G.J., Schimbor, C.M. and May, R.A. (1988) Detection of cytomegalovirus in urine from newborns by polymerase chain reaction DNA amplification. *J. Infect. Dis.* **158:** 1177–1184.
6. Wolf, D.G. and Spector, S.A. (1992) Diagnosis of human cytomegalovirus central nervous system disease in AIDS patients by DNA amplification from cerebrospinal fluid. *J. Infect. Dis.* **166:** 1412–1425.
7. Harris E. *et al.* (1998) Typing of dengue viruses in clinical specimens and mosquitoes by single-tube multiplex reverse transcriptase PCR. *J. Clin. Microbiol.* **36:** 2634–2639.
8. Gingeras, T.R. *et al.* (1991) Use of self-sustained sequence replication amplification reaction to analyze and detect mutations in zidovudine-resistant human immunodeficiency virus. *J. Infect. Dis.* **164:** 1066–1074.
9. Barany, F. (1991) Genetic disease detection and DNA amplification using cloned thermostable DNA ligase. *Proc. Natl. Acad. Sci. USA.* **88:** 189–193.
10. Lomeli, H. *et al.* (1989) Quantitative assays based on the use of replicatable hybridisation probes. *Clin. Chem.* **35:** 1826–1831.
11. Shih, J.W.K., Cheung, L.C., Alter, H.J., Lee, L.M. and Gu, J.R. (1991) Strain analysis of hepatitis B virus on the basis of restriction endonuclease analysis of polymerase chain reaction products. *J. Clin. Microbiol.* **29:** 1640–1641.
12. Boucher, C.A.B. *et al.* (1990) Zidovudine sensitivity of human immunodeficiency viruses from high risk, symptom-free individuals during therapy. *Lancet* **336:** 585–590.
13. Choo, Q.L., Kuo, G and Weiner, A.J. *et al.* (1989) Isolation of a cDNA clone derived from a blood-borne non-A, non-B viral hepatitis. *Science* **244:** 359–362.
14. Matsui, S.M., Kim, J.P. and Greenberg, H.B. *et al.* (1991) The isolation and characterisation of a Norwalk virus-specific cDNA. *J. Clin. Invest.* **87:** 1456–1461.

15. Pilot-Matias, T.J., Muerhoff, A.S. and Simons, J.N. *et al.* (1996) Identification of antigenic regions in the GB hepatitis viruses GBV-A, GBV-B and GBV-C. *J. Med. Virol.* **48:** 329–338.

16. Chang, Y., Cesarman, E., Pessin, M.S., Lee, F., Culpepper, J., Knowles, D.M. and Moore, P.S. (1994) Identification of herpes virus-like DNA sequences in AIDS-associated Kaposi's sarcoma. *Science* **266:** 1865–1869.

17. Nuovo, G.J. *et al.* (1992) Histological distribution of polymerase-chain reaction-amplified human papillomavirus 6 and 11 DNA in penile lesions. *Am. J. Surg. Pathol.* **16:** 269–275.

Chapter 5

Human infectious diseases II – bacteria, fungi, and protozoa

5.1 Introduction

In contrast to its use in virology, molecular methods have made slow progress into the routine diagnosis of other micro-organisms. Many, if not most, of these organisms can be cultivated and phenotypically character-ized. The phenotypic approach has been advanced in some laboratories with semi-automation, such as API or Enterotube systems, but these are a minor part of the routine methodology. Pasteur would feel at home in most hospital and veterinary microbiology laboratories even today. Molecular methods will, however, play an increasing role because of their speed, versatility and specificity. Moreover it is estimated that up to 95% of bacteria present in a sample are noncultivable. Cheaper cultural methods will, no doubt, be used as a standard but there are four areas where nucleic acid based diagnosis will be the method of choice:

- The identification and characterization of fastidious organisms;
- Rapid typing of isolates for epidemiological and other purposes;
- Rapid determination of antimicrobial resistance;
- Identification of organisms in a noncultivable state.

This chapter overviews these aspects with some examples for the different classes of micro-organisms. The emphasis is on medically important infectious agents. The reader is referred elsewhere for detailed methods [1,2]; in addition there are now several peer-reviewed journals devoted to molecular diagnosis of infectious disease and most specialist journals have papers on the subject.

Bacteria are responsible for the majority of currently treatable infec-tious diseases and because of this an accurate rapid diagnosis and deter-mination of susceptibility to antibiotics is important. In addition, typing is important for epidemiology and, ultimately, control of these organisms.

5.2 Specimen collection and preparation

Molecular methods are best applied to purified bacteria but, in practice, this would cause delay in the time to processing of the specimen. Material collected may range from the sterile, relatively acellular, environment of cerebrospinal fluid to the highly complex milieu of food. The more complex the environment, the more likely that there will be substances that interfere with hybridization or inhibit amplification. Complex environments are also likely to have enzymatic activity that degrades nucleic acids. Thus, specimens need to be collected in a manner likely to stabilize nucleic acids for transport to the laboratory but without the addition of materials likely to inhibit subsequent probing. Apart from the special circumstance when differentiation between viable and nonviable bacteria is required, DNA is the target and RNAase inhibition is not usually required. Collection onto ice for rapid transport to the laboratory followed by subsequent phenol-chloroform extraction are standard procedures. Obviously procedures should be as sterile as possible to prevent the addition of exogenous micro-organisms. Further purification before probing or PCR may not be required when bacterial load is likely to be high such as with acute meningitis. If significant nonbacterial contamination is likely, then additional steps to 'purify' the bacterial DNA may be required. Proteinase K digestion, alkaline lysis and thermophilic protease digestion are all commonly used. Water samples, may need filtering to remove large particulates. Soil contains humic substances that inhibit PCR and require separation of bacterial DNA from the sample using resin purification or similar process.

5.3 Identification

The bacterial genome consists of a circular chromosome. In addition there may be extra-chromosomal elements on plasmids which may encode virulence determinants. Five years after Southern described DNA:DNA hybridization detection, Moseley *et al.* applied DNA probes to the detection of *Escherichia coli* (*E. coli*) toxin genes [3]. Since then methods have been developed which facilitate the detection of chromosomal, plasmid or total DNA in a wide range of bacteria. In addition the advent of more advanced genotypic methods, such as PCR, has enabled the detection of previously undiscovered pathogens. Indeed molecular methods have been important in the understanding of pathogenesis and led to a revision of Koch's postulates. The use of probes to 16S ribosomal RNA sequences is peculiar to bacterial diagnostics enabling both detection and phylogenetic analysis.

5.3.1 Mycobacteria

Slow-growing or fastidious bacteria are particularly appropriate targets for molecular diagnostics (*Table 5.1*). The prime example of the former is

Table 5.1: Examples of human pathogenic bacteria that are fastidious or difficult to cultivate *in vitro*

Borrelia spp.
Brucella abortus
Mycobacterium leprae
Mycobacterium tuberculosis
Mycoplasma pneumoniae
Rickettsia spp.
Treponema pallidum
Bacteria in a viable but noncultivable (VNC) state

Mycobacterium tuberculosis. Traditional methods of diagnosis, as with other bacterial pathogens, have relied on microscopy and culture. Microscopic identification with Ziehl-Neelsen (ZN) stain allows a rapid and inexpensive means of identification but is not specific for *M. tuberculosis*. It is also well -recognized to commonly produce false-negative results. Culture on specialized media, such as Lowenstein-Jensen, takes several weeks. Earlier identification of a positive culture can be achieved by radiometric detection of $^{14}CO_2$-labeled palmitic acid released from liquid media, or by activation of fluorescent dyes by CO_2 produced. Neither method is, however, as fast as PCR-based methods.

In the main, PCR-based methods for *M. tuberculosis* utilize probes to either rRNA [4] or to a repetitive DNA sequence, IS6110 [5], although methods based on the detection of other genes, such as those that code for 65kDa [6] and 38kDa [7] proteins have also been described. Utilization of the IS6110 target has an inherent advantage in that, although both *M. tuberculosis* and the related *M. bovis* Bacille-Calmette-Guerin strain are detected, the differences in copy number of the target sequences within the two organisms results in differing patterns of amplification products. Mycobacteria can be found in many bodily fluids but most methods have been developed for use on sputum. Eisenach *et al.* examined 162 sputa for the presence of IS6110 genes. All those that were ZN smear positive were also PCR positive. One of two smear-negative, culture-positive samples was PCR-positive. Two of the 43 smear-negative, culture-negative specimens were also PCR-positive. Twenty six nonmycobacterial specimens were PCR-negative and one of 42 nontuberculous mycobacterial specimens was negative. Although the sensitivity was poor with smear-negative specimens, the sample was small. Other studies have supported the finding, however, that PCR-based methods are more sensitive than ZN stains [8]. Although culture still appears to be more sensitive than single-round PCR, advances such as multiple primer and nested methods should improve sensitivity.

Another mycobacterium, *M. leprae* which causes leprosy, is an example of an organism that cannot be grown routinely (except poorly in armadillo or mouse foot pads). Diagnosis by microscopy has low sensitivity and PCR appears to be both sensitive and specific for fresh-frozen specimens but less so for paraffin-embedded specimens. Other fastidious organisms for which nucleic acid based diagnosis is likely to be the preferred method of detection are shown in *Table 5.1*.

Both these organisms could be cited as examples of organisms that many laboratories would not have the containment facilities to culture but there are many other organisms which would no longer have to be handled in extensive high level containment facilities if they could be inactivated prior to gene diagnosis (*Table 5.2*).

Table 5.2: Examples of human pathogenic bacteria that may require high-level containment facilities for handling in the laboratory

Bacillus anthracis
Brucella abortus
Chlamydia psittaci
Corynebacterium diphtheriae
Coxiella burnetti
Ehrlichia spp.
Francisella tularensis
Mycobacterium tuberculosis
Rickettsia spp.
Salmonella paratyphi
Salmonella typhi
Shigella dysenteriae
Yersinia pestis

5.3.2 Other bacteria

Apart from PCR, the literature describes other amplification methods for the detection of bacteria. The diagnosis of Chlamydia trachomatis in genital and urinary specimens by the probe amplification, ligase chain reaction (LCR) has been shown to be both sensitive and specific. NASBA has been applied to the identification of mycobacteria using universal primers, with subsequent species-typing with specific probes [9]. This method is isothermal and more reliably quantitative than PCR. Strand displacement amplification (SDA) has been described for the detection of *M. tuberculosis*. SDA relies on the ability of certain restriction endonucleases to nick double-stranded DNA. This nick is then extended at the 3' end by Klenow polymerase such that the downstream strand is displaced. This is an isothermal amplification method that is capable of generating a 10^7-fold amplification in 2 hours at 37°C [10].

Many bacteria can exist in both a pathogenic and nonpathogenic state. Merely finding the organism does not imply that is causing the disease. In this scenario, genotypic methods can be used to detect virulence determinants. An example of this would be the use of toxin gene-specific probes to detect toxigenic strains of *Clostridium difficile*, a cause of a particular type of diarrhea associated with the use of antibiotics, termed antibiotic-associated or pseudomembranous colitis [11]. Not all virulence determinants will be chromosomally mediated but probes are equally suited to detect toxin genes on plasmids. *Escherichia coli* is a Gram-negative bacterium which constitutes much of the bacterial flora of the normal human gastrointestinal tract. Certain strains however carry plasmids that code for toxins that cause diarrhea. These plasmids and, even their level of expression, can be detected by specific probes [12]. Probe technology is particularly useful in an environment such as the gut which is loaded with bacteria, and possibly, several toxins which may not be differentiable by phenotypic methods. It almost goes without saying that these genes were first identified by genotypic methods.

5.3.3 Genotypic vs phenotypic

Genotypic methods of identification have the advantage over phenotypic methods of identification such as zymotyping, (identification of variation in enzymatic activity or structure), of being less variable and less dependent on assay conditions. The fact that the genes exist for, for example, a virulence determinant implies that the organism has the capacity to cause damage. Indeed it is now recognised that specific virulence genes are only expressed in the host and not under laboratory conditions [13]. Gene detection is more reproducible than detection of the expressed products or their effects even under differing assay conditions.

The main drawback of most probe methods that are described in the literature is that they do not differentiate between 'live' and 'dead' bacteria. The most common approach to resolve this is to detect mRNA by RT-PCR. This has been successfully applied to the differentiation of, for example, viable from nonviable *Legionella pneumophila* [14]. False positives due to contamination with minute quantities of DNA can be minimized by DNAase digestion of template or by so-called RS-PCR in which the cDNA is tagged with a unique sequence so that only cDNA, produced by the RT step, is subsequently amplified.

5.4 Commercial systems

For purposes other than research, commercially available methods offer 'off the shelf' availability of molecular diagnostics. Probes have been commercially available for several years for organisms such as *Myco-bacterium tuberculosis, Mycoplasma pneumoniae, Neisseria gonorrhoeae* and *Legionella* spp. These have been expensive, technically demanding and often no more rapid and sensitive identification than conventional methods. Their acceptance into routine diagnosis of clinical specimens has been far from widespread. The advent of gene amplification methods has begun to overcome these problems. Most methods are based on PCR for which the world-wide patents are held by Hofmann La Roche, and accrue a royalty payment. These systems (Amplicor®) are semi-automated, based on a PCR-ELISA methodology. Such systems do not require precultivation of organisms. Currently they detect both viable and noncultivable organisms, although it is possible in theory to distinguish between live and dead bacteria. Quantitation, or more accurately, semiquantitation is a feature of the newer available products. Kits are being developed continuously but are already available for M. tuberculosis in sputum. A commercial typing system, EnviroAmp® for *Legionella* species in the environment is also available. This kit uses a set of biotinylated primers to conserved regions of 5S rDNA sequences, then species-specific primers which anneal to the *Mip* gene of *Legionella pneumophila*. Identification of amplicons is by hybridization to membrane bound capture probes and subsequent detection by incubation with streptavidin/horse radish peroxidase conjugate in the presence of the substrate, 33,55 tetramethylbenzidine (TMB). A colorless to blue color change indicates a positive reaction.

LCR (see Section 4.3.2) methods have been developed by Abbott laboratories and have become a commonly used methodology in, for example, the detection of *Chlamydia trachomatis* in urine. This compares favorably with PCR with both methodologies being sensitive to levels beyond those required for clinical relevance [15]. The inherent advantage of LCR is its very high specificity with single base changes resulting in a lack of ligation and subsequent probe amplification.

Transcription-mediated amplification (TMA) systems are also available. These have several advantages compared to PCR and LCR: they can use both RNA and ssDNA targets, increasing sensitivity, and the resulting RNA amplicon is labile so reducing the risk of contamination of subsequent testing. One commercial system (bioMerieux) employs TMA of rRNA followed by a hybridization protection assay (HPA). HPA utilizes acridinium ester-labeled probes to bind to the target amplicon. The addition of a selection reagent results in the hydrolysis of the acridinium ester which releases light detectable by luminometry. Systems are available for the detection of *M. tuberculosis* complex in respiratory secretions and *C. trachomatis* in urethral swab specimens and urine. Developments should enable these kits to be used with automated immunoanalysers, such as the VIDAS system, which are commonly used for serological assays.

5.5 Typing of isolates

5.5.1 Epidemiology

Epidemiological typing of isolates is useful for tracking an outbreak of infection prospectively, and for finding a source retrospectively. The latter is its more common use, in practice, as the aim is to abort the spread of infectious disease. Although it is recognized that within one bacterial species, there may be between 100 and 1000 genetic clones, in clinical practice the source of an outbreak is most frequently a monoclone. Genotypic typing is more discriminatory between related isolates than phenotypic methods: for example, it has been shown that *Staphylococcus aureus* isolates that were identical by zymotype and serotype were distinguishable by genotype, although isolates of an identical genotype were also of the same phenotype [16]. As with their use in identification, there are a number of targets for probes:

- Chromosomal DNA
- Plasmid DNA
- Repetitive DNA

5.5.2 Chromosomal DNA

Chromosomal DNA was the target for an early genotypic typing scheme for bacteria called BRENDA (for Bacterial Restriction ENDonuclease Analysis). Chromosomal DNA is extracted from, ideally, a pure bacterial culture and digested with frequent-cutting restriction enzymes. Subsequent electrophoresis of the fragments results in an RFLP pattern. This technique is not as easily applied to bacteria as it is for some viruses because the much larger genome produces a large number of restriction fragments, some of which may be difficult to resolve on a gel. An example of its use has been the investigation of the source of a laboratory-acquired case of diarrhea due to *Campylobacter jejuni* as a frequently used laboratory strain. Instead of using standard gel electrophoresis, PFGE has also been used as it results in smaller number of bands and an easier to interpret fingerprint. In an outbreak of 32 cases of *Legionella pneumophila*, which causes pneumonia, this technique was used to incriminate a water system as the source. Although genotypic matching was found, phenotyping by monoclonal antibody showed that the patient and water isolates belonged to Pontiac and Bellingham types, respectively [17]. These techniques are dependent on the organism being cultivatable.

Gene amplification techniques are now more commonly employed for the typing of microbial isolates. The first described was arbitrarily-primed PCR (AP-PCR) in which an arbitrary primer is to detect amplification length polymorphisms [18]. This has now been adapted as the RAPD (Random Amplified Polymorphic DNA) method [19]. Both methods have been generically described at DNA amplification fingerprinting (DAF). A short single primer oligonucleotide, five nucleotides or longer, is used so that binding is of low specificity to a large number of loci in the target DNA. These primers are typically of random sequence and greater than 60% G:C base ratio. After PCR amplification, complex banding patterns can be visualized in agarose or polyacrylamide gels. In theory it is possible to obtain a specific fingerprint for an organism but there are some practical difficulties. The use of short primers necessitates low annealing temperatures. Fast ramp rates (the speed with which a PCR machine moves between its different cycles) can then lead to 'melting off' of the primer before an amplification product can be produced. Ramp rates and primer concentrations are crucial to reproducibility, which is often lacking. The target DNA may also be polymorphic, and a single point mutation can lead to a different banding pattern. For large genomes, the banding pattern may be also very complex and be difficult to interpret. The original application of AP-PCR was to identify 24 strains of five staphylococcal species, 11 strains of *Streptococcus pyogenes* and three varieties of the rice plant, *Oryza sativa* [20].

Gene sequence analysis is the ultimate discriminatory tool but until recently has been too labor intensive for routine use. Automated methods using fluorescent chemical markers has made this a more practical proposition.

5.5.3 Plasmid DNA

Plasmid DNA has been recognized, almost since antibiotics were first introduced, as a means of transferring genetic information and, particularly antibiotic resistance, between bacteria. This is regardless of the species of bacterium but is clinically most important amongst Gram-negative bacteria. Analysis of the plasmids that are carried, a plasmid profile, has been used commonly to investigate and characterize outbreaks of infection epidemiologically. The plasmids usually characterized are those that confer antimicrobial resistance, termed R (for resistance) factors. These have been shown to be useful epidemiological markers of the spread of nosocomial infection within hospital settings. The obvious pitfall with using plasmid profiles for epidemiology is that not all bacteria will carry them. In addition, even if present it may be present in such low copy number to be isolatable.

5.5.4 Repetitive DNA

Repetitive DNA is common in bacterial genomes. The identification of the differing frequency of repetitive sequences at differing positions within the chromosomal DNA allow a 'fingerprint' to be established. Chromosomal DNA is digested then hybridized to radioactive, or chemiluminescent, probes. The IS6110 sequence (IS) of *Mycobacterium tuberculosis* is present in low copy number and has been used to in the epidemiology of tuberculosis [21] as well as to identify the organism. Mycobacteria have other IS sequences which have been also been used, for example, to show that mycobacteria sequences found in patients with Crohn's disease had similar IS900 fingerprint patterns to those of *M. paratuberculosis* from cattle and *M. avium* from wood pigeons; this is of interest because both these animals have a bowel disease, caused by the respective mycobacterial species, which is similar to the human condition.

Rep-PCR involves the PCR amplification of regions of the genome between repetitive elements. Differing sizes of product occur as a result of distance variations between the repeat sequences. The repetitive sequences that code for tRNA are conserved and may have variable intervening genome lengths. With judicious use of primer probes to tRNA genes it is then possible to produce DNA fingerprints with multiple bands or single bands of differing size which can be separated by migration rate on agarose gel electrophoresis. It has been possible to type streptococcal and staphylococcal species by this method [22]. The epidemiological analysis of many other Gram-positive and Gram-negative bacteria has been undertaken using rep-PCR.

Ribotyping can be used for epidemiological purposes and is particularly useful for investigating outbreaks due to those bacteria which are not easily typed by phenotypic methods. These include mucoid *Pseudomonas aeruginosa* strains, *Legionella* and *Rhodococcus* spp. Problems still occur, however, because these methods may be too discriminatory.

5.5.5 Phylogeny

Taxonomy requires an identification of similarities as well as differences in a bacterium. Sequence conservation is most commonly sought within the rRNA encoding DNA sequences (rDNA); rRNA genes are found in all bacteria and mutate at a slow rate. Both 5S and 16S ribosomal RNA sequences have been used, the former is smaller and less complex but the major advances have been made using the latter. The use of probes derived from rRNA sequences of *E. coli* that hybridize to total digested bacterial DNA result in a hybridization pattern (usually 7–12) of those chromosome fragments that contain rRNA sequences. The use of rRNA probes for DNA fingerprinting has been termed ribotyping: labeled probes are generally derived from *E. coli* 23S, 16S, and 5S rRNA sequences. Current developments use gene amplification. Primers to conserved regions are used to PCR amplify regions that are then identified using genus-specific and species-specific probes. With the proper design of primers it is possible to amplify all genes of a range of bacteria, so-called universal primers. Such primers have been designed to detect a broad range of bacteria, with subsequent genus and species identification using probes to intervening variable sequences (*Figure 5.1*). In practice, the complexity of the initial amplification with universal primers requires the second specific amplification so that the identification can be made.

The use of 16S rRNA gene probes has enabled the identification of noncultivable bacteria and, indeed, suggests that only a fraction of the total bacterial species have been identified by cultural methods.

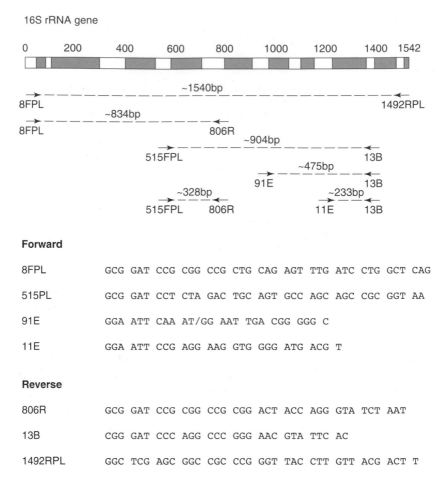

Forward

8FPL   GCG GAT CCG CGG CCG CTG CAG AGT TTG ATC CTG GCT CAG

515PL   GCG GAT CCT CTA GAC TGC AGT GCC AGC AGC CGC GGT AA

91E   GGA ATT CAA AT/GG AAT TGA CGG GGG C

11E   GGA ATT CCG AGG AAG GTG GGG ATG ACG T

Reverse

806R   GCG GAT CCG CGG CCG CGG ACT ACC AGG GTA TCT AAT

13B   CGG GAT CCC AGG CCC GGG AAC GTA TTC AC

1492RPL   GGC TCG AGC GGC CGC CCG GGT TAC CTT GTT ACG ACT T

Figure 5.1: Map of the conserved regions of the 16S rRNA coding region with examples of universal primers (adapted from Relman, 1993).

5.6 Antimicrobial resistance

Conventional determination of antimicrobial resistance consists of culture and identification of the organism and determination of growth in the presence of antibiotic as separate procedures. Gene probes can be used to determine the presence of genetic sequences coding for resistance which is plasmid or chromosomally-mediated; if the latter it may be possible to combine detection of the organism and determination of antibiotic resistance simultaneously. A number of genes responsible for antibiotic resistance have now been identified (*Table 5.3*). One of the best characterized is the *mecA* gene of *Staphylococcus aureus*. This gene encodes a penicillin-binding protein and is present in the majority of methicillin-resistant isolates of *S. aureus* (MRSA), but absent in those that

Table 5.3: Examples of antibiotic resistance genes

Antibiotic resistance genes	Examples of organisms	Antibiotic(s) affected
mecA	*Staphylococcus aureus, S. epidermidis*	Penicillins
rpoB	*Mycobacterium tuberculosis, M. leprae*	Rifampicin
katG	*Mycobacterium tuberculosis*	Isoniazid
tetK	*S. aureus, Streptococcus* spp.	Tetracycline
tetM	*Streptococcus, Staphylococcus, Neisseria, Haemophilus, Mycoplasma, Bacteroides, Bacillus* spp.	tetracycline
ermA	*S. aureus*	Erythromycin
TEM-1	*Neisseria, Haemophilus* spp.	β-lactams
aacC3	*Pseudomonas* spp.	Aminoglycosides

are susceptible. Detection of the *mecA* gene correlates strongly with phenotypic determination of resistance. In one study, in which the *mecA* gene was detected using PCR, 42 of 46 isolates of *S. aureus* and *S. epidermidis* that were *mecA*-positive were also methicillin resistant [22]. Bacterial cells were lysed then genetic material directly probed and amplified within 1 day. Such detection enables appropriate infection control measures and antibiotic therapy, if required, to be instituted more rapidly than by conventional methods. The cost of the methodology is more than outweighed by the potential savings in health resource.

5.7 Novel and noncultivatable bacteria

Perhaps the most exciting application of nucleic acid based diagnosis has been the detection of previously unrecognized organisms either as causes of specific diseases or those that are completely novel. Bacillary angiomatosis is an unusual neoplasia of the microvascular tissue of the skin, most commonly recognized in AIDS patients. Using probes to 16S rRNA sequences it was possible to detect sequences which showed a high degree of homology to *Bartonella quintana*, an organism that caused another disease, bartonellosis [23]. With this knowledge it was possible to, subsequently, isolate Bartonella bacteria from skin lesions of patients with the disease.

Cat scratch disease is a disease predominantly of children and young adults. It typically presents as tender lymphadenopathy following a history of cat scratch with erythema at the site of the scratch. Diagnosis has been conventionally made by the clinical features and characteristic histopathology in the affected lymph node. Since the first description of the disease many viruses and chlamydiae have been implicated as etiological agents. The observation that bacilli seen in cat scratch disease had a striking similarity morphologically to those seen in bacillary angiomatosis led to a search for the same agent, *Bartonella henselae*, in the latter condition . Several studies using a variety of techniques, but particularly PCR, have now established this as, at least, the main causative agent.

An even more dramatic use of molecular technology has been its application to the discovery of the agent of Whipple's disease. This is a multisystem disease with malabsorption as a central component. As with bacillary angiomatosis and cat scratch disease, bacteria-like organisms had been seen in affected tissues but were not cultivable. By amplification and quantification of 16S rDNA sequences it was possible to show that there were too many bacteria present in lesions to represent simple colonization. By sequence analysis of amplified rDNA sequences it was also possible to identify a new organism. This organism has been provisionally named *Tropheryma whippelii* and is probably an actinomycete [24]. Currently, it is still not possible to cultivate this organism and nucleic acid based diagnosis using specific primers is the only means of detecting the organism.

5.8 Fungi

The majority of the common fungal causes of human diseases can be rapidly diagnosed by Gram stain or potassium hydroxide microscopy and culture. This is cheap and sensitive and would remain the method of choice for candidoses and dermatophyte infections. Even with chronic infections, such as zygomycoses, microscopy of infected material enables a diagnosis within minutes, compared to the hours that a molecular probe method would take. For disorders such as meningitis due to *Cryptococcus neoformans*, even though microscopy is often insensitive, the added use of latex agglutination for cryptococcal polysaccharide allows for a rapid and reliable diagnosis.

Molecular probes will, however, have a limited role in diagnostic mycology. Life-threatening infections due to fungi are becoming more prevalent because of acquired immunosuppression due to AIDS and transplantation procedures. Rapid diagnosis of disseminated candidiasis and invasive aspergillosis could be improved by molecular methods. Currently diagnosis depends on culture and/or serology. Both of these are insensitive, slow or both. The use of antigen detection tests speeds up the process but has less than 100% sensitivity or specificity, neither exceeding 70% in practice.

Although direct probe methods have been described, as with the rest of diagnostic microbiology, amplification methods are more likely to be used in a routine setting. Highly conserved ribosomal DNA (rDNA) targets have been used as nongenus-specific targets for PCR amplification [25]. The use of PCR to detect genus-specific targets, such as the chitin synthase gene, have been shown to detect as few as 10 organisms in blood and other samples within 6 hours [26]. PCR detection can also be used to monitor therapy. The use of 5S rDNA probes in a PCR amplification assay to detect *Pneumocystis carinii* in sputum samples from AIDS patients was used to monitor treatment with pentamidine. The shedding patterns correlated consistently with clinical outcome even though the organisms were not reliably detected by cytology [27].

Any of the probe technologies that have been described for viruses and bacteria could be applied to the detection of fungi. There are some problems which have particular relevance to the application of such methods to the mycological diagnosis. The first is that with the serious acute infections, blood is often the most appropriate sample. Blood contains a number of inhibitors of PCR, such as hemoglobin. Secondly, fungal cell walls are hardy structures and resist release of nucleic acid targets. Most methods described also incorporate initial centrifugation and washing steps to clean up the DNA that is released. The third aspect to be considered in the interpretation of molecular assays is that fungi are very common environmental contaminants and indeed may be present in specimens without probable clinical significance. The specificity of PCR for the detection of *Pneumocystis carinii* is high but its sensitivity is also high, higher than immunofluoresence, and in one study the organism was detectable in the absence of clinical symptoms [28].

The use of probes for the epidemiology of fungal infections is not as widespread as that for bacteria but the same methods could be applied in an identical manner. Fungal DNA contains several DNA repeat sequences suitable for fingerprinting.

5.9 Protozoa

Apart from the clinical and epidemiological features, the diagnosis of protozoal infection is conventionally made by microscopy or serology. Culture is a specialized procedure which requires animal house facilities. When parasite load is high, microscopy is an adequate methodology for the diagnosis of acute infection. Serodiagnosis is useful for screening purposes but often lacks specificity as cross-reactive antigens are common. Nucleic acid probes offer the advantages of sensitivity and speed.

The most frequent protozoal infection world-wide is malaria which is caused by one of four *Plasmodium species*. Probes, when allied to amplification, have advantages over conventional microscopy when low-level parasitemia occurs and when it is useful to identify drug-resistant mutants. In the vast majority of situations, microscopy will remain the method of choice. Probes to rRNA genes can be genus- or species-specific. They can also be employed to detect mixed infections which are not infrequent. Oligonucleotide rRNA probes are able to detect down to 10–50 parasites in blood spots [29]. PCR detection enables even greater sensitivity than routine light microscopy. Probes have also been described which detect gene sequences coding for circumsporozoite protein, a 21 bp repetitive sequence, and a number of parasite specific genes. PCR assays have been developed that probe for specific mutations in the dihydrofolate reductase gene that confer resistance to the antifolate drugs (such as proguanil and pyrimethamine). These assays seem to be clinically useful [30].

Although probe assays are described in the literature for a number of other protozoal pathogens (including *Trichomonas vaginalis, Cryptosporidium parvum, Naegleria fowleri, Toxoplasma* spp., and *Giardia lamblia*) it is unlikely that these will supersede conventional methods as they are cumbersome and expensive. Exceptions might however occur in specific situations. *Entamoeba histolytica* has recently been recognized as being two species, one pathogenic which retains the name and the other nonpathogenic which has been renamed *E. dispar*. These are morphologically identical but one causes diarrhea and invasive amebiasis whereas the other does not. They can be distinguished by differing isoenzyme patterns or by probes. Probes to two highly abundant repetitive sequences, one 145 bp (P-145) the other 133 bp (B-133), the 125kDa surface antigen gene, and the small subunit rRNA enable differentiation of pathogenic *E. histolytica* from nonpathogenic species. PCR assays may also become routine in the public health setting such as the detection of *Cryptosporidium* and *Naegleria* spp. in water samples where conventional methods are often too insensitive.

Techniques such as RAPD-PCR can also be applied to understand the epidemiology of these organisms. Protozoa have hypervariable mini-satellite tandem repeat sequences which can be exploited by mini-satellite variant repeat PCR (MVR-PCR) for epidemiological purposes [31].

5.10 Future prospects

Nucleic acid probes are becoming a standard diagnostic methodology in microbiology. Evaluation studies are already showing that for many micro-organisms they are cost-effective when compared to standard cultural and serological methods. Automation is the key to the high throughput and rapid turnover of specimens that will be required. This will not only lead to a more specific and sensitive detection of micro-organisms but drive down the financial costs of molecular diagnostics. The next major advance will be the incorporation of DNA chip technology into routine diagnostics. Probes are bound to a solid phase for detection of unknown, 'interrogated,' DNA. The use of a solid phase to bind probes is not novel, being used in commercial PCR systems, but DNA chips involve a very large array of probes of known sequence on a solid surface, 'the chip.' This technology has already been used, with hexamer probes, for DNA sequencing. It has also been applied, empirically, to detect mutations in the HIV-1 protease gene, with 12 224 probes of differing sequence bound to a glass surface [32]. PCR amplicons were labeled with fluorescein then applied to the probe array. An accuracy of 98–99%, when compared to conventional sequencing, was obtained. With the increasing recognition that most bacteria are yet to be recognised, DNA chip technology, allied to sequencing, may speed up the process of establishing true bacterial biodiversity.

Biosensor technology to detect binding of bound probe also offers the prospect of rapid throughput and automation. One possibility is the use of membranes that respond chromatically to target-probe binding, for example, the diacetylenic lipids. Their use has already been described for the detection of bacterial toxins [33].

References

1. Persing, D.H., Smith, T.H., Tenover, F.C. and White, T.J. (eds). (1993) *Diagnostic Molecular Microbiology, Principles, and Applications*. American Society for Microbiology pp. 1–641.
2. Ehrlich, G.D. and Greenberg, S.J. (eds). (1994) *PCR-based Diagnostics in Infectious Disease*. Blackwell, Boston pp. 1–698.
3. Moseley, S.L., Huq, I., Alim, A., So, M., Samadpour-Motalebi, M. and Falkow, S. (1980) Detection of enterotoxigenic *Escherichia coli* by DNA colony hybridization. *J. Infect. Dis.* **142:** 892–898.
4. Boddinghaus, B., Rogall, T., Flohr, T., Blocker, H. and Bottger, E.C. (1990) Detection and amplification of mycobacteria by amplification of rRNA. *J. Clin. Microbiol.* **28:** 1751–1759.
5. Eisenach, K.D., Cave, M.D., Bates, J.H. and Crawford, J.T. (1990) Polymerase chain reaction amplification of a repetitive DNA sequence specific for *Mycobacterium tuberculosis. J. Infect. Dis.* **161:** 977–981.
6. Hance, A.J., Grandchamp, B. and Levy-Frebault, V. *et al.* Detection and identification of mycobacteria by amplification of mycobacterial DNA. *Mol. Microbiol.* **3:** 843–849.
7. Sjobring, U., Meckelburg, M., Andersen, A.B. and Mjorner, H. (1990) Polymerase chain reaction for detection of *Mycobacterium tuberculosis. J. Clin. Microbiol. 28:* 2200–2204.
8. Savic, B., Sjobring, U., Alugupalli, S., Larsson, L. and Mjorner, H. (1992) Evaluation of polymerase chain reaction, tuberculostearic acid analysis, and direct microscopy for the detection of Mycobacterium tuberculosis. *J. Infect. Dis.* **166:** 1177-1180.
9. Van der Vliet, G., Schukkink, R.A.F., van Gemen, B., Schepers, P. and Klatser, PR. (1993) Nucleic acid sequence based amplification (NASBA) for the identification of mycobacteria. *J. Gen. Microbiol.* **139:** 2423–2429.
10. Walker, G.T., Little, M.C., Nadeau, J.G. and Shank, D.D. (1992) Isothermal in vitro amplification of DNA by a restriction enzyme/DNA polymerase system. *Proc. Natl. Acad. Sci. USA* **89:** 392–396.
11. Gumerlock, P.H., Tang, Y.J., Weiss, J.B. and Silva J. (1993) Specific detection of toxigenic strians of Clostridium difficile in stool specimens. *J. Clin. Microbiol.* **31:** 507–511.
12. Gyles, C., So, M. and Falkow, S. (1974) The enterotoxin plasmids of *Escherichia coli. J. Infect. Dis.* **130:** 40–49.
13. Mahan, M.J., Slauch, J.M. and Mekalanos, J.J. (1993) Selection of bacterial

virulence genes that are specifically induced in host tissues. *Science* **259:** 686–688.

14. Mahbubani, M.H., Miller, R.D., Atlas, R.M., Dicesare, J.L. and Haff, L.A. (1991) Detection of bacterial mRNA using polymerase chain reaction. *Biotechniques* **10:** 48–49.

15. Dille, B.J., Butzen, C.C. and Birkenmeyer, L.G. (1993) Amplification of *Chlamydia trachomatis* DNA by Ligase Chain Reaction. *J. Clin. Microbiol.* **31:** 729–731.

16. Schlichting, C., Branger, C. and Fournier, J.-M. *et al.* (1993) Typing of Staphylococcus aureus by pulsed field gel electrophoresis, zymotyping, capsular typing, and phage typing: resolution of clonal relationships. *J. Clin. Microbiol.* **31:** 227–232.

17. Struelens, M.J. et al. (1992) Genotypic and phenotypic methods for investigation of a nosocomial *Legionella pneumophila* outbreak and efficacy of control measures. *J. Infect. Dis.* **166:** 22–30.

18. Welsh, J. and McClelland, M. (1990) Fingerprinting genomes using PCR with arbitrary primers. *Nucl. Acids Res.* **18:** 7213–7218.

19. Williams, J., Kubelick, A.R., Livak, K.J., Rafalski, J.A. and Tingey, S. (1990) DNA polymorphisms amplified by arbitrary primers are useful as genetic markers. *Nucl. Acids Res.* **18:** 6531–6535.

20. Welsh, J. and McClelland, M. (1991) Genomic fingerprints produced by PCR with consensus tRNA gene primers. *Nucleic Acids Res.* **19:** 861–866.

21. Hermans, P.W.M., van Soolingen, D. and Dale, J.W. *et al.* (1990) Insertion element IS986 from *Mycobacterium tuberculosis*: a useful tool for diagnosis and epidemiology of tuberculosis. *J. Clin. Microbiol.* **28:** 2051–2058.

22. Unal, S., Hoskins, J. and Flokowitsch, J.E. *et al.* (1992) Detection of methicillin-resistant staphylococci by using the polymerase chain reaction. *J. Clin. Microbiol.* 1992; **30:** 1685–1691.

23. Relman, D.A., Loutit, J.S., Schmidt, T.M., Falkow, S. and Tompkins, L.S. (1990) The agent of bacillary angiomatosis. *N. Engl. J. Med.* **323:** 1573–1580.

24. Relman, D.A., Shmidt, T.M., MacDermott, R.P. and Falkow, S. (1992) Identification of the uncultured bacillus of Whipple's disease. *N. Eng. J. Med.* **327:** 293–301.

25. Makimura, K., Muramaya, S.Y. and Yamaguchi, H. (1994) Detection of a wide range of fungi in clinical specimens using polymerase chain reaction (PCR) amplification. *J. Med. Microbiol.* **40:** 358–364.

26. Buchman, T.G., Rossier, M., Merz, W.G. and Charache, P. (1990) Detection of surgical pathogens by *in vitro* DNA amplification. Part 1. Rapid identification of *Candida albicans* by *in vitro* amplification of a fungus-specific gene. *Surgery* **108:** 38–347.

27. Oka, S., Kitada, K., Kohjin, T., Nakamura, Y., Kimura, S. and Shimada, K. (1993) Direct monitoring as well as sensitive diagnosis of *Pneumocystis carinii* pneumonia by the polymerase chain reaction on sputum samples. *Mol. Cell. Probes* **7:** 419–424.

28. Elvin, K. (1994) Laboratory diagnosis and occurrence of *Pneumocystis carinii*. *Scand. J. infect. Dis.* **94:** 1–34.

29. Waters, A.P. and McCutchan, T.F. (1989) Rapid sensitive diagnosis of malaria based on ribosomal RNA. *Lancet* **I:** 1343–1346.

30. Zolg, J.W., Chen, G.X. and Plitt, J.R. (1990) Detection of pyrimethamine resistance in *Plasmodium falciparum* by mutation specific polymerase chain reaction. *Mol. Biochem. Parasitol.* **39:** 257–266.

31. Arnot, D.E., Roper, C. and Sultan, A.A. (1994) MVR-PCR analysis of hypervariable DNA sequences variation. *Parasitol. Today* **10:** 324–327.
32. Kozal, M.J., Shah, N. and Shen, N. *et al.* (1996) Extensive polymorphisms observed in HIV-1 clade B protease gene using high-density oligonucleotide arrays. *Nature Med.* **2:** 753–759.
33. Charych, D., Cheng, Q. and Reichert, A. *et al.* A 'litmus test' for molecular recognition using artificial membranes. *Chem. and Biol.* **3:** 113–120.

Chapter 6

Applications of nucleic acid diagnosis in pathology

6.1 Introduction

Nucleic acid-based methods to improve diagnosis of many diseases are currently being introduced into routine pathology departments. Histo-pathology is based upon macroscopic and microscopic examination of tissue biopsies, resections or autopsy specimens to determine the pathological processes taking place and diagnose the disease or cause of death. However, this morphological approach has many limitations and although in the majority of cases it allows pathologists to classify the specimen it does not always identify the biological basis of the disease which may be needed for successful therapy. The application of nucleic acid diagnostic methods to pathology will improve our understanding of the molecular mechanisms causing human disease and provide a means of screening for individuals at risk. Many of the early molecular events that lead to disease take place before the clinical symptoms develop. Therefore the application of sensitive methods that can detect these changes in cells from body fluids and smaller less invasive biopsies will allow treatment to be used to prevent the progression of the disease.

6.2 Which pathological processes can we detect by nucleic acid methods?

This chapter will identify the current applications of nucleic acid-based methods in pathology, which are centered round the diagnosis of tumors and infectious agents. In both situations a novel nucleic acid target is present allowing the disease to be detected. Although these methods can be applied to extracted nucleic acids for an accurate diagnosis it is often necessary to identify the cells carrying the target nucleic acids using *in situ* techniques. We are now also able to monitor the progression of other nonneoplastic diseases such as degenerative changes, autoimmune reactions and adaptations by measuring gene expression in tissue biopsies. Changes in the profile of mRNA expression in these tissues can be diagnostic and also indicate the severity, stage of the disease and any response to treatment. In this chapter it would be impossible to cover thoroughly all the applications of nucleic acid based diagnosis in pathology therefore I have focused on techniques used to diagnosis tumors.

6.3 Tumor diagnosis

6.3.1 The genetic basis of cancer

Before describing the applications of nucleic acid methods in the diagnosis of human tumors it is necessary to describe the background to the genetic basis of cancer. The full background details to this field have recently been extensively reviewed including the inherited cancer syndromes [1,2]. Today we clearly accept the genetic basis of cancer and this is supported by evidence that cancer or the risk of cancer can be inherited, that mutagens cause tumors in humans and that tumors are monoclonal in origin. Monoclonality depends on a clonal evolution of tumor cells therefore the cells of the tumor all show the same genetic characteristics of the original transformed cell. Studies of carcinogenesis have shown that the majority of tumours develop through a multistage process where each step requires the activation, mutation or loss of specific genes. Early genetic changes are needed to block molecular pathways leading to programmed cell death and to maintain the cell's ability to proliferate without terminal differentiation. This transformed state can then further evolve into an invasive tumor cell, which can control its blood supply by stimulating new blood vessel growth a process called angiogenesis, spread to other tissues and organs by a process called metastasis and develop resistance to chemotherapy and radiotherapy. The genes that play a major role in the early development of cancer can be divided into two classes, the oncogenes and the tumor-suppressor genes.

6.3.2 Oncogenes

Oncogenes were first identified through the study of acute transforming avian or rodent retroviruses. They represent homologues of mammalian genes that have been incorporated into the viral genome and in these circumstances they can transform the host cell. Their normal functioning counterparts called proto-oncogenes have important roles in controlling many diverse signalling pathways within the cell including embryonic development, cell proliferation, differentiation and programed cell death. Although these transforming retroviruses have not been implicated in human cancer, many of the oncogenes identified in these animal viruses are present in human cancers and are activated by mutagenic events that convert the proto-oncogene into an active oncogene. Mutations in the proto-oncogenes in cancer cells such as point mutations, chromosomal translocations, and gene amplifications convert the gene by either mutating the protein product to an active form or increasing the level of expression of the wild-type protein. These somatic mutations represent a dominant genetic change to the pre-neoplastic or neoplastic cell, as they require a mutation in only one allele. Over 50 oncogenes have been identified in human cancer but not all of these are present in animal retroviruses. *Table 6.1* gives a list of some of the more important human oncogenes and their diagnostic potential.

6.3.3 Tumor-suppressor genes

In contrast to oncogenes, which are single allele mutations, tumor-suppressor genes are dependent on mutation of both alleles so both chromosomal copies of the gene are affected and are therefore characterized by loss of function of the target gene. These genes were first identified from mapping familial cancer genes such as the retinoblastoma (*RB1*) gene and sequencing the mutations. Knudson [3] proposed a two hit model for the development of these familial pediatric cancers. In the familial form one mutant allele is inherited there-fore increasing the risk of the development of retinoblastoma by muta-tion or loss of the wild-type allele. Sporadic cases are less likely, as mutation of both alleles in the same cell would have a lower probability. Like oncogenes tumor-suppressor genes also have diverse functions influencing tumor cell proliferation, differentiation and death but a further group also influences DNA repair and maintain genome stability [4]. This group of genes has a more indirect role in the development of cancer as their inactivation leads to an increased risk of further mutagenesis of other oncogenes or tumor-suppressor genes. *Table 6.2* gives a list of tumor-suppressor genes in human cancer and their importance to diagnostic pathology.

6.4 Nucleic acid-based methods for the detection and diagnosis of cancer

The mutations and chromosomal aberrations that are found in tumor cells are detected in clinical samples using a variety of methods to analyze the nucleic acids. These mutations include point mutations, truncations, amplifications, and translocations characteristic of oncogene activation and point mutations, deletions, loss of heterozygosity, microsatellite instability and promoter hypermethylation, which are characteristic of tumor-suppressor genes.

Table 6.1: A list of some of the more important human oncogenes and their diagnostic potential

Oncogene	Tumor type	Activation mechanism	Diagnostic potential	Reference
K-Ras	Carcinomas	Point mutation of p21 GTPase	Marker of Lung cancer in bronchoalveolar lavage fluid	[48]
N-Ras	Myeloid leukemia	Point mutation of p21 GTPase	13% case positive in acute myeloid leukemia gives better prognosis	[49, 50]
H-Ras	Bladder	Point mutation of p21 GTPase	10% cases positive can be detected in urine sediment	[51, 52]
EGFR	Gliomas carcinomas	Amplification of growth factor receptor gene	40% of cases positive but not present in benign disease	[53, 54]
NEU	Breast and other carcinomas	Amplification of growth factor receptor gene	19% of case positive poor prognosis and shorter disease free survival	[55, 56]
c-MYC	Burkitt's lymphoma SCLC and other carcinomas	Translocation or amplification of transcription factor gene	3%-30% of cases frequency increases with treatment	[57]
N-MYC	Neuroblastoma, SCLC	Amplification of transcription factor gene	10% of Neuroblastoma show amplification most relevant adverse prognostic factor	[58]
L-MYC	SCLC	Amplification of transcription factor gene	25% of primary SCLC show amplification of one *myc* gene associated with poor prognosis and progression	[59, 60]
BCL-2	Non-Hodgkin's B-cell lymphoma	Translocation of antiapoptotic gene	BCL-2 expression rather than 14:18 translocation important to prognosis	[61, 62]
CYCD1	Breast and other carcinomas, B-cell lymphoma, parathyroid adenoma	Translocation or amplification of cyclin D gene	10% of cases show amplification not associated with clinical features	[63]
BCR-ABL	Chronic myeloid leukemia, acute myeloid leukemia (T-cell)	Translocation of chimeric non-receptor kinase gene deregulating tyrosine-kinase activity	RT-PCR of chimeric mRNA used for detection of minimal residual disease	[64, 65]
RET	Thyroid Papillary or medullary type (MEN2)	Translocation or point mutation of transmembrane growth neurotrophic receptor with tyrosine kinase gene	Mutation detection allows prophylactic thyroidectomy to be performed in RET mutation-positive patients at an earlier stage of the disease process	[66, 67]

Table 6.1: *continued*

Oncogene	Tumor type	Activation mechanism	Diagnostic potential	Reference
CDK4	Sarcoma (familial) Gilomas	Amplification or point mutation of cyclin dependent kinase gene	Approximately 10% of sarcomas show amplification of this oncogene. Linked to disease progression rather than survival	[68, 69]
MET	Renal carcinoma papillary type (hereditary)	Point mutation of Hepatocyte growth factor receptor tyrosine kinase gene	Inherited and somatic forms found in high grade carcinomas	[70]
SMOH	Basal cell carcinoma	Point mutations of Smoothened gene in sonic hedgehog pathway	Important in some cases of BCC	[71]
HSTF1	Gastric esophageal cancers	Amplification of FGF like growth factor	50% amplified in esophageal cancers	[72]
PML-RARα	Acute promyelocytic leukemia	Translocation to chimeric retinoic acid receptor transcription factor	Important diagnostic and prognostic information in the mangement of APL patients	[73]
E2A-PBX	Pre B acute lymphocytic leukemia	Translocation chimeric transcription factor	Detects 95% of childhood ALL but no prognostic information	[74]
MDM2	Sarcoma	Amplification of P53 binding protein	see CDK4	[68]

Table 6.2: A list of tumor suppressor genes in human cancer and their importance to diagnostic pathology (based on Fearon 1997 [2])

Gene	Protein product	Sporadic tumor types	Familial tumors	Diagnostic potential	Reference
RB1	Cell cycle regulator that binds transcription factors	Retinoblastoma, adenocarcinoma, SCLC breast prostate and bladder	Retinoblastoma	Different clinical presentation with LOH	[75]
TP53	Transcription factor	Approximately 50% of all cancers but <10% in NHL, prostate neuroblastoma	Li-Fraumeni syndrome	Often associated with very poor prognosis	[76, 77, 78]
P16	Cyclin dependent kinase inhibitor	Significant in many cancers	Melanoma and pancreatic carcinoma	$p16$ gene is genetically inactivated in 98% of pancreatic carcinomas through different mechanisms	[79]
APC	Regulates β-catenin function by microtubule binding	Colorectal tumors	Familial adenomatous polyposis (FAP)	FAP represents 1% of colorectal tumors while 30% show LOH at the APC locus	[80, 81]
MSH2 MLH1P MS1 PMS2	DNA mismatch repair	Many cancers "mutator phenotype" e.g. colorectal, gastric endometrial	Hereditary non polyposis colorectal cancer (HNPCC)	HNPCC represents 5% of colorectal tumors while 30 show high replication error rate RER+ tumors	[82, 83]
NF1	P21ras GTPase microtubule binding	Melanoma neuroblastoma	Neurofibromatosis type 1	Expression of protein diagnostic	[84]
NF2	Membrane link to cytoskeleton	schwannomas	Neurofibromatosis type 1	Expression of protein diagnostic	[84]
VHL	Regulator of protein stability	Hemangioblastomas, renal carcinoma, phaeochromocytoma	Von Hippel-Lindau (VHL) disease	Alellic variation affects clinic outcome	[85]
BRCA1 BRCA2	DNA repair with HRAD51	Ovarian rare in breast cancer	inherited ovarian and breast cancer	Large number of mutations low sporadic frequency	[86]
PTCH	Transmembrane receptor for sonic Hedghog pathway regulator of smoothened protein	Basal cell carinoma medulosarcoma	Gorlin syndrome	Gene mRNA expression upregulated in sporadic and familial BCC	[87]

Table 6.2: *continued*

Gene	Protein product	Sporadic tumor types	Familial tumors	Diagnostic potential	Reference
pTEN	Tyrosine phosphatase	Gliomas breast, prostate cancer	Cowden's disease	Rarely found in sporadic prostate cancer	[88, 89]
DCC	netrin-1 receptor component	Colorectal and other cancers	Not known	Biological marker of progression	[88]
DPC4	Downstream signalling in TGFβ pathway	Pancreatic and other cancers	Not known	Associated with LOH on chromsome 18	[91]
E-CAD	Transmembrane cell cell adhesion molecule	Diffuse gastric, lobular breast and others	Not known	Mutations are involved in the the etiology of sporadic lobular breast cancers	[92]
β-CAT	Transcriptional activator links E-cadherin to cytoskeleton	Melanoma, colorectal	Not known	Mutations are rare in invasive colorectal cancers	[93]
TGFβRII	Transmembrane receptor for TGFβ	Colorectal, gastric and others	Not known	Mutations in right-sided colorectal tumors containing RII mutations resemble those with the RER+ phenotype	[82]

6.4.1 DNA extraction from tumor cells

DNA is a stable molecule in these samples as long as the tissues are suitably treated to prevent nuclease digestion. If the tissue samples are fixed using aldehyde fixatives such as formalin and embedded in paraffin wax, the DNA is chemically modified and degraded into small fragments of average size about 500 bp. However, even these degraded fragments of DNA can be amplified by PCR to analyze many genetic mutations.

One major obstacle to the analysis of somatic mutation in tumor tissue is the presence of abundant nonneoplastic cells in the specimen. This nonneoplastic background often includes cells from the stroma, blood vessels, and immune response and may mask genetic changes that otherwise would be easily detectable in a pure tumor cell population. The problem may be overcome by using microdissection techniques to sample small populations of cells from a tissue section [5]. This allows genetic analysis using DNA samples extracted from pure tumor or a background cell population [6]. The technique can be applied to frozen and paraffin-embedded tissue allowing the analysis of archive clinical samples.

Tumor cell populations can also be isolated from blood or other body fluids using monoclonal antibodies attached to magnetic beads [7,8]. This system can capture small numbers of cells selectively and the extracted DNA can be analyzed by PCR. The technique provides a noninvasive method for the early detection of disseminating disease. PCR methods can therefore be applied to DNA extracted from very small numbers of tumor cells or nonneoplastic cells to provide germ-line comparisons. Tumor DNA can also be detected in human plasma, urine and feces and used to diagnose the presence of a tumor this DNA is thought to be released from tumor cells undergoing cell death and survives as short fragments of degraded DNA which can be amplified by PCR. By detecting known genetic changes which are common to many tumors it is possible to identify tumor cells as the source of this DNA and confirm the presence of a malignancy. Nontumor DNA may be required from the patient to make these comparisons [9].

6.4.2 Detection of mutations

The detection of specific point mutations in oncogenes and tumor-suppressor genes in clinical samples has until recently been limited by the technology. A variety of methods are currently available to either identify mutations at previously determined sites or screen the entire gene sequence for the presence of base mutations at an unknown site. Most of these techniques are based upon PCR amplification of the target sequence followed by gel analysis. They include allele specific amplification (see ARMS in Section 3.4.1), SSCP analysis [10,11], HA [12] (see Section 2.4.3), base mismatch recognition [13], semi-automated DNA sequence analysis [14] (see Section 2.4.6) and micro-array hybridization sequencing [15] (see below). Base mismatch recognition uses bacteriophage resolvases T4 endonuclease VII and T7 endonuclease I for detecting mutations in genomic DNA. Heteroduplex DNA fragments prepared by amplification from DNA containing known mutations are cleaved by one or both enzymes at nucleotide mismatches. The digested DNA fragments are analyzed by gel electrophoresis.

Using semi-automated sequencing techniques a single case can take several weeks to fully sequence the coding region of a typical oncogene or tumor-suppressor gene from genomic DNA. mRNA can be a better source for sequencing by conversion to cDNA using reverse transcriptase. However, the preservation of mRNA in clinical samples can be a problem and even cDNA sequencing to screen for mutations is time consuming.

An alternative approach is to use micro-array hybridization sequencing to rapidly detect mutations in oncogene or tumor-suppressor genes [16]. In this technique sequence analysis of DNA is carried out by hybridization of RNA transcripts with oligonucleotide array microchips, for example Genechips. The RNA transcripts of PCR-amplified genomic DNA are fluorescent labeled by enzymatic or chemical methods and hybridized to the microchips. The simultaneous measurement in real time of the hybridization and melting on the entire oligonucleotide array is carried out with a fluorescence microscope equipped with CCD camera. Choosing the proper allele-specific oligonucleotides from among the set of overlapping oligomers optimises the microchip diagnostics by computer analysis. In comparison with automated sequencing the accuracy of mutation detection can be increased by simultaneous hybridization of the microchip with two differently labeled samples and by parallel monitoring of their hybridization with a multiwavelength fluorescence microscope [15]. A recent comparison of this rapid and high throughput method with standard automated sequencing showed at least an equal accuracy and confirmed the diagnostic potential for this technique [17].

6.4.3 Detection of translocations

Recently many chromosomal translocations involved in human neoplasia have been defined at the molecular level. In addition to advancing the understanding of pathological mechanisms underlying the transformation process, the cloning and sequencing of the genes altered by the translocations have allowed PCR primers to be developed to detect the translocation for diagnosis and monitoring of patients. In particular, PCR methodology yields rapid, sensitive and accurate diagnostic and prognostic information. The specific translocation sequence can also be used diagnostically to provide early detection of minimal residual disease (MRD) [18]. Examples of translocations that are currently used to monitor MRD include: follicular lymphoma with t(14;18) [19,20] (see

Section 6.5 investigating lymphoma), chronic myeloid leukemia and acute lymphoblastic leukemia (ALL) with t(9;22), ALL with t(4;11), and acute myeloid leukemia (AML) with t(8;21) or t(15;17) [21]. Molecular assays for specific gene fusions also provide a genetic approach to the differential diagnosis of soft tissue sarcomas. The genetic categories correspond closely to the standard histopathologic categories. Therefore, the RT-PCR assays for chimeric mRNA transcripts are used for an objective assessment of paediatric soft tissue sarcomas [22].

6.4.4 Comparative genomic hybridization

CGH is based on a two-color, competitive fluorescence *in situ* hybridization of differentially labeled tumor and reference DNA to normal metaphase chromosomes (see Section 2.5.3). This technology has made a great impact in molecular pathology because it can be applied to archival specimens to create copy number karyotypes throughout the whole genome from very small amounts of DNA. If chromosomal imbalances can be correlated with etiological and clinical features of tumors, CGH could be able to provide new prognostic and diagnostic criteria [23]. CGH findings further provide starting points for the molecular genetic characterization of altered chromosomal regions harboring yet unidentified genes involved in tumorigenesis and tumor progression [24].

6.4.5 Microsatellite analysis

In addition to specific mutations in oncogenes and tumor suppressor genes, changes in the DNA repeat sequences, called microsatellites (see Section 3.2), can also be used to detect the clonal evolution of neoplastic cells. Microsatellites are tandem repeats of simple di-, tri-, or tetranucleotides and can be amplified by PCR using flanking primers. They are highly polymorphic and interspersed throughout the genome and therefore can be used to analyze specific loci for loss of heterozygosity (LOH). The microsatellite analysis can be used to investigate tumors for genetic instability, demonstrated as the increase or reduction of the repeat elements, which might have resulted from the disruption of mismatch repair genes such as hMSH2 and hMLH1 [25,26].

Microsatellite DNA markers have been widely used as a tool for the detection of loss of heterozygosity and genomic instability in primary tumors. For example in a recent study, urine samples from patients with suspicious bladder lesions that had been identified cystoscopically were analyzed by this molecular method and by conventional cytology. Microsatellite changes matching those in the tumor were detected in the urine sediment of 95% of patients who were diagnosed with bladder cancer, whereas urine cytology detected cancer cells in 50% of the samples. These results suggest that microsatellite analysis, which in principle can be performed at about one-third the cost of cytology, may be a useful addition to current screening methods for detecting bladder cancer [27].

6.4.6 Hypermethylation of the tumor suppressor gene promoter

Tumor cells also show widespread genomic hypomethylation, regional areas of hypermethylation, and increased DNA-methyltransferase (DNA-MTase) activity. This methylation imbalance is believed to contribute to tumor progression where genomic hypomethylation is probably due to cell transformation and novel gene activation. The regional hypermethylation sites are normally unmethylated CpG islands located in gene promoter regions. This hypermethylation correlates with transcriptional inhibition that can serve as an alternative to promoter or coding region mutations for the inactivation of tumor-suppressor genes. Examples of these genes include specific inhibitors of cyclin-dependent kinases p16INK4a [28] and p15INK4b [29], retinoblastoma (Rb) [30], and E-cadherin but hypermethylation is not the only mechanism to inactivate these suppressor genes nor is it confined to specific tumors. The process of regional hypermethylation evolves during tumorgenesis where the protection of unmethylated CpG islands is lost possibly by chronic exposure to increased DNA-MTase activity and disruption of local protective mechanisms. Hypermethylation of some genes appears to occur after the onset of neoplastic transformation while other genes, such as the estrogen receptor, become hypermethylated in normal cells during ageing. Thus hypermethylation may predispose malignant transformation in some tumors [31].

The methylation of specific bases can be assessed using methylation sensitive restriction enzymes. These restriction sites in the gene promoter are tested by digesting the tumor DNA with methylation sensitive and insensitive enzymes followed by Southern blotting analysis or PCR amplification across the restriction site. The methylation status of a tumor-suppressor gene is assessed using several restriction sites in CpG sequences flanking the 5' and 3' region of the gene. The identification of CpG hypermethylation in the gene promoter should correlate with loss of mRNA synthesis in the tumor cells and the methylation status of the gene in normal tissue [32].

6.4.7 Telomerase activity

For tumor cells to become immortal the number of telomeric repeat sequences must be maintained allowing DNA polymerase priming at the telomeres of each chromosome. The ribonucleoprotein telomerase is the enzyme which copies these repeats and is expressed by most malignant tumors. This enzyme is inactive in normal somatic cells except for male germ cells and proliferating stem cells. Thus, the measurement of telomerase activity in tissue samples may provide useful diagnostic and prognostic information. Telomerase activity is measured using a modified, semiquantitative PCR-based telomeric repeat amplification protocol (TRAP). The assay requires extraction of the enzyme activity from frozen clinical samples. This is followed by primer-directed PCR amplification of a telomere extension reaction. The telomerase activity in tumor samples has been shown to be modified by inhibitors in some specimens [33]. Therefore, the true activity may be difficult to determine. Studies carried out on various types of tumor suggest that most tumors benign or malignant are positive for telomerase enzyme although the amount of active enzyme tends to be higher in malignant cases [34].

6.4.8 Viruses associated with human cancer

Most viral infections produce a lytic phase and therefore usually induces an immune response to eliminate the viral damage. In latent or chronic infections, the viral genome can be incorporated into the human genome or is maintained as episomal sequences. This type of viral infection can lead to the development of cancer as the expression of viral genes can regulate normal cell growth, proliferation, differentiation and apoptosis. These oncogenic viruses act as promoters by immortalizing the host cell, and increasing the probability of its malignant transformation. The viral oncogenic proteins interfere with host cell tumor-suppressor gene functions causing malignant transformation in some infected cells. Alternatively, some viruses carry genes that behave like oncogenes which when activated may also lead to malignant transformation.

Development of human cancers, both benign and malignant, has been linked to several viruses. Unlike many animal tumors most human cancers are not associated with known retroviruses. Human retroviruses such as HIV and human T-cell leukemia/lymphoma virus (HTLV) have been linked to non-Hodgkin's lymphoma but these tumors may arise as a result of immunosuppresion rather than a direct viral oncogenic mechanisms.

The main viruses involved in human cancers include HPV, hepatitis B virus (HBV) and several Human Herpes viruses. *Table 6.3* gives a list of these viruses, tumors, viral genes, and their diagnostic potential. Of the Herpes Viruses, Epstein-Barr herpes virus (EBV), (human herpesvirus-4) is associated with oral hairy leukoplakia, lymphoproliferative disease, lymphoepithelial carcinoma, B and T cell lymphomas such as Burkitt's lymphoma, Hodgkin's disease, and nasopharyngeal carcinoma. Human herpesvirus-8 has been implicated in all forms of Kaposi's sarcoma, primary effusion lymphomas, multiple myeloma, angioimmunoblastic lymphadenopathy, and Castleman's disease. Human herpesvirus-6 has been detected in lymphoproliferative disease, non-Hodgkin's lymphomas, Hodgkin's disease, and oral squamous cell carcinoma. These Herpes

Table 6.3: A list of oncogenic viruses, associated tumors, viral genes and their diagnostic potential

Oncogenic virus	Tumors	Viral oncogene	Diagnostic potential	Reference
Human immunodeficiency virus (HIV)	Indirect cause of many tumors through immune suppression during AIDS	None	Risk of cancer with AIDS	[94]
Human T-cell leukemia/ lymphoma virus (HTLV)	T-cell leukemia/lymphoma	Tax oncoprotein of HTLV1 chronically activates transcription factor NF-kappaB	Detection of provirus in lymphoma cells for diagnosis	[95]
Simian virus 40 (SV40)	Human mesotheliomas	SV40 large T-antigen (Tag), inactivates growth suppressive proteins such as the retinoblastoma family and p53	Detection of viral DNA in tumors and viral gene expression	[96, 97]
Human papilloma virus (HPV)	Cutaneous and mucosal squamous carcinomas	Viral types 16, 18, 31 and 45 have viral oncogenes E6 and E7	Detection of particular HPV types could be used in the diagnosis and management of cervical cancer	[35]
Hepatitis B virus (HBV) and HCV virus	Hepatocellular carcinoma (HCC)	HBX gene expression associated with HCC	Detection of HBV and HCV	[98]
Epstein-Barr virus	Lymphoproliferative diseases, Hodgkin's disease, Burkitt's lymphoma, nasopharyngeal carcinoma, gastric carcinoma	BHRF1 behaves like antiapoptosis LMP-1 interacts with signalling from CD40/CD40-L, which promotes growth in B cells. EBNA-2 mimics the Notch 1 pathway	Detection of lytic and latent EBV genes expressed in malignant cells	[99, 100]
Human Herpes virus 8	Karposi's sarcoma, primary effusion lymphomas, multiple myeloma, angioimmunoblastic lymphadenopathy, and Castleman's disease	HHV8-V interleukin like 6 expression also HHV8-Vcyc, with sequence similarity to human G1 cyclins	Detection of key V-oncogenic gene expression in malignant cells	[101, 102]
Human Herpes virus 6	Lymphoproliferative disease, non Hodkin's lymphomas, Hodgkin's disease, and oral squamous cell carcinoma	HHV-6-ORF-1 is an oncogene that binds to and affects p53	Detection of key V-oncogenic gene expression in malignant cells also integration sites	[103]

viruses cause widespread asymptomatic infection and are associated with the malignant transformation in only small number of infected individuals. Therefore other carcinogenic factors must influence the development of these viral associated tumors. For example SV40 infected human populations via the polio vaccines that were distributed to millions of people from 1955 through early 1963. This virus has now been implicated in the etiology of several human tumors including mesothelioma where it probably acts as a cocarcinogen with asbestos fibres.

The detection of human viruses that are associated with the development of human cancer is now based upon PCR/RT-PCR detection of the viral genome or RNA expressed from viral genes. The design of primers that are viral specific can be a problem as viral sequences are often homologous with human genes thus care must be taken to detect regions of the viral genome that are specific to the viral strain under investigation. The use of *in situ* hybridization can also be important to confirm that the viral genes are present in the tumor cells and RNA *in situ* hybridization can demonstrate viral oncogene expression in the tumor.

6.4.9 Human papilloma virus and cervical screening

HPV is a good example where DNA based diagnostic tests may have a role to identify early and premalignant changes [35]. The role of human papillomavirus in benign (squamous papilloma, focal epithelial hyperplasia, condyloma acuminatum, verruca vulgaris), premalignant (oral epithelial dysplasia), and malignant (squamous cell carcinoma) neoplasms within the oral cavity is well recognized. HPVs can be classified biologically or phylogenetically into cutaneous or mucosal types. Cutaneous papillomaviruses produce benign skin tumors (warts) which occur commonly on the hands, face and feet. They spread readily among children and young adults during recreational activities. Exposure to sunlight sometimes causes these lesions to progress to skin cancer.

HPV has also been strongly associated with the development of cancer at mucosal sites one of the most important is cervical carcinoma. These HPV strains are the most common sexually transmitted virus, infecting both men and women and are transmitted from the vagina at birth. Genital infection usually clears within a few months, but may persist in some individuals. HPV types vary and are related to the degree of cervical dysplasia present. HPV types 6 and 11 are are common and are associated with low-grade squamous epithelial lesions where as HPV types 16, 18, 31, and 45 are less common but are present in cervical intraepithelial neoplasia which can lead to invasive cancer.

Although visible lesions are present in some cases infected with HPV, the majority of individuals with HPV genital tract infection do not have clinically apparent disease. Therefore screening could be used to identify individuals with active or chronic infection. Commercial dot blot hybridization, DNA-RNA hybrid capture assays and PCR analysis are available for laboratory diagnosis of genital HPV. Detection of particular HPV types could be used in the diagnosis and management of cervical cancer, for resolving borderline cytology, to distinguish between high-grade and low-grade disease and to improve the accuracy of routine cervical screening [36].

6.4.10 *Detection of* Helicobactor pylori *in gastric carcinoma*

The multistep carcinogenesis of gastric carcinoma is complex and involves alterations in oncogenes, tumor suppressor genes, telomerase activity as well as genetic instability. Several recent epidemiological studies have demonstrated a close association between *Helicobacter pylori* infection and carcinoma of the mid- or distal-stomach. Several mechanisms have been proposed by which *H. pylori* infection might lead to predisposition for gastric cancer. These include mechanisms, such as increased gastric epithelial proliferation in response to *H. pylori*, lowered gastric ascorbic acid levels, and high occurrences of atrophic gastritis. However, little evidence has emerged to support a direct mechanism. *H. pylori*-associated inflammation may interact with other causal factors related to gastric carcinogenesis and can result in the intestinal type of gastric cancer. In this model of *H. pylori* induced carcinogenesis DNA damage due to oxygen radicals induced by persistent inflammatory cell infiltrations in the gastric mucosa may lead to genetic mutations and result in the development of diffuse-type carcinoma. PCR has been used extensively to investigate the role of different strains of [37], associated virulence genes [38] and frequency of infection in gastric juices [39], and biopsies [40].

6.4.11 *Confirming tumor clonality*

Methods to analyze the clonality of tumor cell populations can be very useful clinically. Markers of clonality can be used to aid primary diagnosis, to study the clonal evolution of the tumor and to detect signs of early relapse. Tumors with specific chromosomal abnormalities such as a translocation, inversions, rearrangements and mutations can be tested for clonality by showing that all the tumor cells carry these genetic aberrations. Alternatively, where no clear genetic alteration can be detected analysis of X-linked gene polymorphisms can be used, such as the phosphoglycerate kinase (*PGK*), hypoxanthine phosphoribosyl transferase (*HPRT*), monoamine oxidase A (*MAOA*) genes, the CAG repeat of the human androgen receptor (*HU-MARA*) gene, and the hypervariable DXS255 gene [41,42]. The method relies on digestion of DNA within the target gene using methylation-sensitive restriction enzymes such as *SnaBI*, PCR amplification of the target gene, and detection of a restriction or microsatellite polymorphism. With this approach, only the inactive (methylated) allele is selectively amplified by PCR. In a polyclonal population both alleles will be present because methylation will affect each allele at random but in clonal cell populations only one allele will be recognized. The accuracy of this approach depends upon the proportion of tumor cell DNA in the sample. Background normal cells will provide polyclonal signal therefore the use of microdissection to enrich for tumor cells improves the reliability of this method which can also be applied to DNA isolated from archival formalin-fixed tissue. Problems can occur where the methylation of the X chromosome is not random and skewed distributions have been demonstrated in normal cell populations. However, the major disadvantage with this technique for detecting clonality is that it limited to the analysis of biopsies from female patients and can only be applied to cases with an appropriate gene polymorphism.

The clonality of non-Hodgkin's lymphomas can be shown by PCR analysis of immunoglobulin antigen receptor and T cell receptor gene rearrangements. These genes rearrange by somatic recombination during B and T cell differentiation and this process is required to provide the diversity of antibody and T cell receptor molecules allowing the immune system to respond to infections when challenged. Therefore each gene rearrangement can be used to identify a clone of B or T cells and tumor populations derived from these cells. PCR amplification of the specific sites of recombination using primers to conserved sequences in the immunoglobulin antigen receptor and T cell receptor genes can be analysed by polyacrylamide gel electrophoresis and SSCP or HA to identify size and sequence differences. Polyclonal populations show a range of bands representing the diversity of gene rearrangement present in the cells while clonal populations are identified from unique bands representing a single allele or both alleles rearranged (see Section 6.5, investigating lymphoma).

6.4.12 Gene expression studies

Oncogenic and tumor suppressor gene changes in tumor cell populations can be used as markers of malignant change to characterize the tumor type, confirm the diagnosis and investigate the tumor phenotype. Other genes that are not directly mutated during tumor cell transformation but are expressed in the tumor can also be informative. These genes may be used to identify the origin of tumor cell, its metastatic potential and drug resistance. For example in breast cancer the expression of a functional estrogen receptor (ER) by the tumor cells would indicate that the tumor may respond succesfully to antiestrogen therapy. Therefore investigating the expression of genes regulated by the ER would determine the ER status of the tumor. To investigate gene expression in cell populations mRNA has to be extracted from a tumor biopsy or mRNA *in situ* hybridization is required to localize the site of expression in the tumor. Both techniques require the mRNA to be preserved in the tissue by rapid tissue processing. Either the tissue must be rapidly frozen immediately after excision or the RNase enzymes must be inhibited by RNA stabilization solution or for *in situ* studies by fixation using formalin based fixatives.

One of the best available methods of evaluating synthetic rates in human tissue biopsies is to measure the amount of a specific mRNA as the amount of mRNA coding for a specific protein is assumed to reflect the rate at which the cell is making that protein. Although there are exceptions to this assumption, this approach has been widely accepted. Unfortunately the methods used to detect small amounts of mRNA have some limitations. Traditional methods require microgram amounts of total RNA extracted from large biopsies but may not represent the specific cellular population under study. *In situ* hybridization gives good localization, but it cannot detect molecules which are present in very low copy number and it does not give consistent quantitation. Reverse transcription of the mRNA to cDNA, followed by use of the polymerase chain reaction (RT-PCR), is more sensitive and can give better quantitation, but does not provide localization. The development of in situ PCR methods may in due course provide excellent localization, but quantitation is unlikely to be precise.

In view of the ability of PCR to detect specific sequences in tiny samples, one strategy has been the microdissection of tiny areas of tissue, often from tissue sections, to provide a relatively well defined starting sample [43]. Quantitation of this approach using a competitive templates has been sufficient to reproducibly detect small changes of mRNA levels [44]. Therefore although mRNA can be isolated from large tissue biopsies or resected tumors more accurate information can be obtained using small microdissected areas [45] or even single isolated cells [46].

mRNA *in situ* hybridization can be used when the cellular site of gene expression is crucial to diagnosis. One example of this is the detection of light chain immunoglobulin in Hodgkin's disease to demonstrate that the Reed-Sternberg cells are monoclonal [47]. This can be determined if these cells all express the same light chain immunoglobulin gene as polyclonal populations of B-cells express both kappa and lambda light chains. In this case the location of gene expression is needed as the RS cells represent only a small proportion of the tumor population which is made up of other inflammatory cells inducing B-lymphocytes and plasma cells.

6.5 Investigating lymphoma

This section provides an example of how nucleic acid methods are used to improve diagnosis. The diagnosis of non-Hodgkin's lymphoma is based on histological and immunocytochemical methods to determine clonality and cell lineage based on the detection of T and B cell antigens on the surface of cells. Clonality often can be assessed in B cell malignancies with the presence of either kappa or lambda light chains on the surface of the neoplastic cells. In other B or T cell lymphoid malignancies clonality is less easily determined. Occasionally, the presence of mixed cell populations or lack of detectable surface antigens may lead to nondiagnostic or ambiguous results. Since the T cell receptor or the immunoglobulin genes rearrange in a unique way for each lymphocyte during lymphocyte development, molecular techniques can be used for the detection of clonal populations of T or B cells.

When the immunoglobulin heavy chain gene rearranges, a V (variable), D (diversity), and J (joining) region are brought together to form a functional unit. While V-D-J joining occurs, random numbers of DNA bases are inserted between the V-D and D-J junctions, forming the third complementarity determining region (Frame CDRIII) that is involved in antigen specificity. Consequently, all nonneoplastic B cells have slightly different distances between the V and J segments. (see *Figure 6.1*)

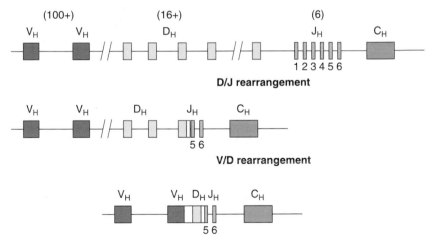

Figure 6.1: Immunoglobulin gene rearrangements combining of Variable V_H Diversity D_H and Joining J_H regions. Two separate recombination events allow the expression of a novel antibody heavy chain. The first event allows one of the 16+ Diversity D_H regions to recombine with one of the 6 Joining J_H regions. The second event allows one of the 100+ Variable V_H to recombine with the previously selected Diversity D_H.

6.6 Southern blotting technique to detect gene rearrangements

DNA is extracted from frozen tissue or liquid specimens (peripheral blood, bone marrow aspirate, body fluids), digested with restriction endonucleases, separated by agarose gel electrophoresis, transferred to a membrane (the Southern blot technique), and then hybridised with ^{32}P labeled probes to Immunoglobulin or T-Cell receptor genes. Clonal rearrangements are detected as DNA fragments migrating differently from germline DNA restriction fragments (*Figure 6.2*). This technique is able to detect a clonal population comprising as few as 1-5% of the total cells in the specimen.

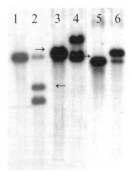

Figure 6.2: Southern blotting of immunoglobulin heavy chain gene for control lymph node (tracks, 1, 3, and 5) and lymphoma tissue DNA (tracks 2, 4, and 6) each sample was digested with HindIII, EcoRI and BamHI restriction endonucleases respectively. Germline bands (arrowed) are shown in control tracks, 1, 3, and 5. Rearranged bands are shown in tracks 2, 4, and 6 for lymphoma DNA sample confirming monoclonality.

6.6.1 B cell clonality detection by the polymerase chain reaction (IgH PCR)

PCR allows amplification of target DNA, the ends of which are determined by the primers. For B cell clonality analysis, two PCR primers are used corresponding to a consensus sequence common to most V segments, and a consensus sequence common to most J segments. When the DNA is amplified, the length of the PCR product is determined by the number of random nucleotides added at the time of VDJ joining (see *Figure 6.3*)

In a nonneoplastic population of B cells, each PCR product is slightly different in size. When these products are separated by electrophoresis, a smear is seen on the gel. In the case of B cell lymphoma, all the neoplastic cells have the same VDJ rearrangement. Therefore, the neoplastic cells all produce a PCR product of the same size, resulting in the appearance of a discrete band on the gel (see *Figure 6.4*).

The PCR method has a number of advantages over Southern blotting technique including the ability to use less DNA, shorter turnaround time, and the ability to analyze paraffin-embedded specimens. B cell clonality

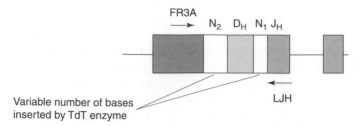

Figure 6.3: PCR amplification of immunoglobulin heavy chain gene using conserved primers to frame 3A of the variable region V_H FR3A and the leading joining region LJH. The size of the amplified product depends upon the number of bases N1 and N2 inserted by the terminal deoxynucleotidyl tranferase enzyme (TdT) during immunoglobulin heavy chain gene rearrangements. DNA extracted from polyclonal populations of B-cells will show a wide range of sizes for this amplicon as the insertions up to 100 bases are possible. DNA from monoclonal populations will show one or two amplified bands corresponding to each allele rearranged.

PCR detection of Ig heavy chain gene rearrangements

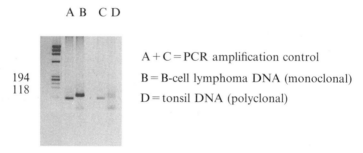

A + C = PCR amplification control

B = B-cell lymphoma DNA (monoclonal)

D = tonsil DNA (polyclonal)

Figure 6.4: PCR product detection by horizontal gel eletrophoresis showing that B-cell lymphoma has a single discrete band (track B) compared to the smear of DNA for the tonsil sample (track D).

analysis can be performed on small specimens such as endoscopic biopsies that often yield insufficient DNA for Southern blot analysis. The ability to perform gene rearrangement studies on fixed tissue can result in substantial cost savings for patients who in the past would have had repeat biopsies to obtain fresh or frozen tissue for gene rearrangement studies when frozen tissue was not saved initially.

6.7 Detection of translocations

6.7.1 *BCL-2 Translocation analysis by the polymerase chain reaction (BCL-2 PCR)*

Greater than 85% of follicular lymphomas have a translocation in which the BCL-2 gene on chromosome 18 is translocated into the immunoglobulin heavy chain locus on chromosome 14. This translocation can be detected by PCR using primers specific for both major and minor translocation breakpoints on chromosome 18, and a second primer specific for chromosome 14 sequences. If t(14;18) is present, a PCR product is produced (see *Figures 6.5 and 6.6*). If the translocation is not present, no product is produced. Since a negative result could indicate DNA that cannot be amplified by PCR, control reactions amplifying the BCL-2 region are performed to verify the quality of the DNA. This technique is capable of detecting as few as one in 100 000 cells, and thus can be used to detect the

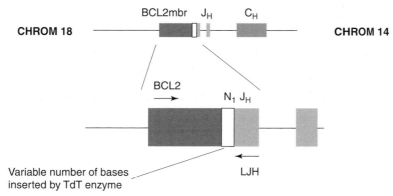

Figure 6.5: Diagram to show the PCR primers used to detect the BCL-2-immunoglobulin heavy chain gene translocation (14:18). Primers are located in the BCL-2 gene and the conserved Joining J_H region of the immunoglobulin gene. The use of BCL-2 primers for the major and minor breakpoints improve the PCR detection efficiency of this translocation.

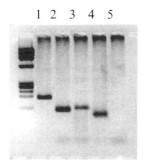

Figure 6.6: An example of PCR amplification of the bcl-2-Ig heavy chain gene (14:18) ranslocation using oligonucleotide primers for the conserved immunoglobulin heavy chain joining region, the heavy chain variable region and bcl-2 gene major and minor break points on DNA from a case of follicular lymphoma. Track 1, control amplification of Bcl-2 major breakpoint. Track 2, control amplification of Bcl-2 minor breakpoint. Track 3, amplification of immunoglobulin gene rearrangement to confirm monocolonality. Track 4, amplification of Bcl-2 major-Ig heavy chain-joining showing translocation. Track 5, amplification of Bcl-2 minor-Ig heavy chain joining.

presence of residual tumor cells in blood or bone marrow following therapy (often termed minimal residual disease). The test can also be performed to confirm the diagnosis of follicular lymphoma in cases where classification is difficult. As with IgH PCR, this technique can be performed both on fresh/frozen or fixed tissue specimens.

6.8 Future developments

The future of nucleic acid based diagnosis in pathology looks very promising with the discovery of more disease-associated genes and the role of genetic factors in influencing disease risk. We will be able to monitor individuals with these risks and detect tumors and other pathologies earlier giving a better chance successful therapy. Advances in genome analysis coupled with vast amounts of genetic data resulting from the Human Genome Project will broadening the scope of research and provide tools to identify individuals at increased risk of disease. The molecular characterization of disease will also allow more successful therapies to be developed which will be more biological in their action and potentially less toxic to the patient compared to current therapies.

References

1. Haber, D.A. and Fearon, E.R. (1998) The promise of cancer genetics. *Lancet* **351 suppl. 2:** SII1–8.
2. Fearon, E.R. (1997) Human cancer syndromes: clues to the origin and nature of cancer. *Science* **278**(5340): 1043–1050.
3. Knudson, A.G.J. (1971) Mutation and Cancer: a statistical study of retinoblastoma. *Proc. Natl. Acad. Sci. USA*, **68:** 820–823.
4. Kolodner, R.D. (1995) Mismatch repair: mechanisms and relationship to cancer susceptibility. *Trends Biochem. Sci.* **20**(10): 397–401.
5. Moskaluk, C.A. and Kern, S.E. (1997) Microdissection and polymerase chain reaction amplification of genomic DNA from histological tissue sections. *Am. J. Pathol.* **150**(5): 1547–1552.
6. Zhuang, Z. and Vortmeyer, A.O. (1998) Applications of tissue microdissection in cancer genetics. *Cell. Vis.* **5**(1): 43–48.
7. Denis, M.G. *et al.* (1997) Detection of disseminated tumor cells in peripheral blood of colorectal cancer patients. *Int. J. Cancer* **74**(5): 540–544.
8. Makarovskiy, A.N. *et al.* (1997) Application of immunomagnetic beads in combination with RT-PCR for the detection of circulating prostate cancer cells. *J. Clin. Lab. Anal.* **11**(6): 346–350.
9. Nollau, P., Moser, C. and Wagener, C. (1996) Isolation of DNA from stool and bodily fluids for PCR amplification. *Biotechniques* **20**(5): 784–788.
10. Sheffield, V.C. *et al.* (1993) The sensitivity of single-strand conformation polymorphism analysis for the detection of single base substitutions. *Genomics* **16**(2): 325–332.
11. Gelfi, C. *et al.* (1997) Detection of p53 point mutations by double-gradient, denaturing gradient gel electrophoresis. *Electrophoresis* **18**(15): 2921–2927.
12. Glavac, D. and Dean, M. (1995) Applications of heteroduplex analysis for mutation detection in disease genes. *Hum. Mutat.* **6**(4): 281–287.

13. Mashal, R.D., Koontz, J. and Sklar, J. (1995) Detection of mutations by cleavage of DNA heteroduplexes with bacteriophage resolvases. *Nat. Genet.* **9(2):** 177–183.
14. Murphy, M. *et al.* (1997) TP53 mutation in ovarian carcinoma. *Eur. J. Cancer* **33(8):** 1281–1283.
15. Hacia, J.G. *et al.* (1996) Detection of heterozygous mutations in BRCA1 using high density oligonucleotide arrays and two-colour fluorescence analysis [see comments]. *Nat. Genet.* **14(4):** 441–447.
16. Kononen, J. *et al.* (1998) Tissue microarrays for high-throughput molecular profiling of tumor specimens [In Process Citation]. *Nat. Med.* **4(7):** 844–847.
17. Gunthard, H.F. *et al.* (1998) Comparative performance of high-density oligonucleotide sequencing and dideoxynucleotide sequencing of HIV type 1 pol from clinical samples. *AIDS Res. Hum. Retroviruses* **14(10):** 869–876.
18. Gribben, J.G. and Nadler, L.M. (1994) Detection of minimal residual disease in patients with lymphomas using the polymerase chain reaction. *Important Adv. Oncol.* 117–129.
19. Crescenzi, M. *et al.* (1988) Thermostable DNA polymerase chain amplification of t(14;18) chromosome breakpoints and detection of minimal residual disease. *Proc. Natl. Acad. Sci. USA* **85(13):** 4869–4873.
20. Fey, M.F. *et al.* (1991) The polymerase chain reaction: a new tool for the detection of minimal residual disease in haematological malignancies. *Eur. J. Cancer* **27(1):** 89–94.
21. Drexler, H.G. *et al.* (1995) Recurrent chromosomal translocations and fusion genes in leukemia- lymphoma cell lines. *Leukemia* **9(3):** 480–500.
22. Barr, F.G. *et al.* (1995) Molecular assays for chromosomal translocations in the diagnosis of pediatric soft tissue sarcomas. *Jama* **273(7):** 553–557.
23. Lapierre, J.M. *et al.* (1998) Comparative genomic hybridization: technical development and cytogenetic aspects for routine use in clinical laboratories. *Ann. Genet.* **41(1):** 56–62.
24. Zitzelsberger, H. *et al.* (1997) Comparative genomic hybridisation for the analysis of chromosomal imbalances in solid tumours and haematological malignancies. *Histochem. Cell Biol.* **108(4–5):** 403–417.
25. Marra, G. and Boland, C.R. (1996) DNA repair and colorectal cancer. *Gastroentero.l Clin. North Am.* **25(4):** 755–772.
26. Habano, W., Sugai, T. and Nakamura, S. (1998) Mismatch repair deficiency leads to a unique mode of colorectal tumorigenesis characterized by intratumoral heterogeneity. *Oncogene* **16(10):** 1259–1265.
27. Mao, L. *et al.* (1996) Molecular detection of primary bladder cancer by microsatellite analysis. *Science* **271(5249):** 659–662.
28. Kashiwabara, K. *et al.* (1998) Correlation between methylation status of the p16/CDKN2 gene and the expression of p16 and Rb proteins in primary non-small cell lung cancers. *Int. J. Cancer* **79(3):** 215–220.
29. Drexler, H.G. (1998) Review of alterations of the cyclin-dependent kinase inhibitor INK4 family genes p15, p16, p18 and p19 in human leukemia-lymphoma cells. *Leukemia* **12(6):** 845–859.
30. Stirzaker, C. *et al.* (1997) Extensive DNA methylation spanning the Rb promoter in retinoblastoma tumors. *Cancer Res.* **57(11):** 2229–2237.
31. Baylin, S.B. *et al.* (1998) Alterations in DNA methylation: a fundamental aspect of neoplasia. *Adv. Cancer Res.* **72:** 141–196.
32. Vonlanthen, S. *et al.* (1998) Expression of p16INK4a/p16alpha and p19ARF/p16beta is frequently altered in non-small cell lung cancer and correlates with p53 overexpression. *Oncogene* **17(21):** 2779–2785.

33. Sugino, T. *et al.* (1997) Telomerase activity and its inhibition in benign and malignant breast lesions. *J. Pathol.* **183(1):** 57–61.
34. Yashima, K. et al. (1997) Telomerase activity and in situ telomerase RNA expression in malignant and non-malignant lymph nodes. *J. Clin. Pathol.* **50(2):** 110–117.
35. Swygart, C. (1997) Human papillomavirus: disease and laboratory diagnosis. *Br. J. Biomed. Sci.* **54(4):** 299–303.
36. Trofatter, K.F., Jr (1997) Diagnosis of human papillomavirus genital tract infection. *Am. J. Med.* **102(5A):** 21–17.
37. Miehlke, S. *et al.* (1999) Molecular relationships of Helicobacter pylori strains in a family with gastroduodenal disease. *Am. J. Gastroenterol.* **94(2):** 364–368.
38. Queiroz, D.M. *et al.* (1998) cagA-positive Helicobacter pylori and risk for developing gastric carcinoma in Brazil. *Int. J. Cancer* **78(2):** 135–139.
39. Monti, J. *et al.* (1998) [Helicobacter pylori detection by polymerase chain reaction in gastric juice and its correlation with the histology (Giemsa)]. *Acta Gastroenterol. Latinoam.* **28(5):** 335–336.
40. Tomtitchong, P. *et al.* (1998) Helicobacter pylori infection in the remnant stomach after gastrectomy: with special reference to the difference between Billroth I and II anastomoses. *J. Clin. Gastroenterol.* **27(suppl 1):** S154–158.
41. Peng, H. *et al.* (1997) Clonality analysis in tumours of women by PCR amplification of X-linked genes. *J. Pathol.* **181(2):** 223–227.
42. Hotta, T. (1997) Clonality in hematopoietic disorders. *Int. J. Hematol.* **66(4):** 403–412.
43. Bicknell, G.R. *et al.* (1996) Amplification of specific mRNA from a single human renal glomerulus, with an approach to the separation of epithelial cell mRNA. *J. Pathol.* **180(2):** 188–193.
44. Hall, L.L. *et al.* (1998) Reproducibility in the quantification of mRNA levels by RT-PCR-ELISA and RT competitive-PCR-ELISA. *Biotechniques* **24(4):** 652–658.
45. Hiller, T. Snell, L. and Watson, P.H. (1996) Microdissection RT-PCR analysis of gene expression in pathologically defined frozen tissue sections. *Biotechniques* **21(1):** 38–40, 42, 44.
46. Trumper, L.H. *et al.* (1993) Single-cell analysis of Hodgkin and Reed-Sternberg cells: molecular heterogeneity of gene expression and p53 mutations. *Blood* **81(11):** 3097–3115.
47. Hell, K. *et al.* (1993) Demonstration of light chain mRNA in Hodgkin's disease. *J. Pathol.* **171(2):** 137–143.
48. Scott, F.M. *et al.* (1997) High frequency of K-ras codon 12 mutations in bronchoalveolar lavage fluid of patients at high risk for second primary lung cancer. *Clin. Cancer Res.* **3(3):** 479–482.
49. Neubauer, A. *et al.* (1994) Prognostic importance of mutations in the ras proto-oncogenes in de novo acute myeloid leukemia. *Blood* **83(6):** 1603–1611.
50. Coghlan, D.W. et al. (1994) The incidence and prognostic significance of mutations in codon 13 of the N-ras gene in acute myeloid leukemia. *Leukemia* **8(10):** 1682–1687.
51. Saito, S. *et al.* (1997) Screening of H-ras gene point mutations in 50 cases of bladder carcinoma. *Int. J. Urol.* **4(2):** 178–185.
52. Cristaudo, A. *et al.* (1997) A simple method to reveal possible ras mutations in DNA of urinary sediment cells. *J. Environ. Pathol. Toxicol. Oncol.* **16(2–3):** 201–204.
53. Collins, V.P. (1995) Gene amplification in human gliomas. *Glia* **15(3):** 289–296.

54. Schober, R. *et al.* (1995) The epidermal growth factor receptor in glioblastoma: genomic amplification, protein expression, and patient survival data in a therapeutic trial. *Clin. Neuropathol.* **14(3):** 169–174.

55. Berns, E.M. *et al.* (1992) Prevalence of amplification of the oncogenes c-myc, HER2/neu, and int-2 in one thousand human breast tumours: correlation with steroid receptors. *Eur. J. Cancer* **28(2–3):** 697–700.

56. Mansour, E.G., Ravdin, P.M. and Dressler, L. (1994) Prognostic factors in early breast carcinoma. *Cancer* **74(suppl. 1):** 381–400.

57. Brennan, J. *et al.* (1991) myc family DNA amplification in 107 tumors and tumor cell lines from patients with small cell lung cancer treated with different combination chemotherapy regimens. *Cancer Res.* **51(6):** 1708–1712.

58. Rubie, H. *et al.* (1997) N-Myc gene amplification is a major prognostic factor in localized neuroblastoma: results of the French NBL 90 study. Neuroblastoma Study Group of the Societe Francaise d'Oncologie Pediatrique. *J. Clin. Oncol.* **15(3):** 1171–1182.

59. Makela, T.P., Saksela, K. and Alitalo, K. (1992) Amplification and rearrangement of L-myc in human small-cell lung cancer. *Mutat. Res.* **276(3):** 307–315.

60. Noguchi, M. *et al.* (1990) Heterogenous amplification of myc family oncogenes in small cell lung carcinoma. *Cancer* **66(10):** 2053–2058.

61. Johnson, A. et al. (1995) Incidence and prognostic significance of t(14;18) translocation in follicle center cell lymphoma of low and high grade. A report from southern Sweden. *Ann. Oncol.* **6(8):** 789–794.

62. Martinka, M. *et al.* (1997) Prognostic significance of t(14;18) and bcl-2 gene expression in follicular small cleaved cell lymphoma and diffuse large cell lymphoma. *Clin. Invest. Med.* **20(6):** 364–370.

63. Seshadri, R. *et al.* (1996) Cyclin DI amplification is not associated with reduced overall survival in primary breast cancer but may predict early relapse in patients with features of good prognosis. *Clin. Cancer Res.* **2(7):** 1177–1184.

64. Hochhaus, A. et al. (1996) Monitoring the efficiency of interferon-alpha therapy in chronic myelogenous leukemia (CML) patients by competitive polymerase chain reaction. *Leukemia* **11 suppl 3:** 541–544.

65. Mar-Aguilar, F. *et al.* (1998) Detecting residual bcr-abl transcripts in chronic myeloid leukaemia patients using coupled reverse transcriptase-polymerase chain reaction with rTth DNA polymerase. *Clin. Lab. Haematol.* **20(4):** 221–224.

66. Learoyd, D.L. *et al.* (1997) Genetic testing for familial cancer. Consequences of RET proto-oncogene mutation analysis in multiple endocrine neoplasia, type 2 [see comments]. *Arch. Surg.* **132(9):** 1022–1025.

67. Lam, A.K. *et al.* (1998) Ret oncogene activation in papillary thyroid carcinoma: prevalence and implication on the histological parameters. *Hum. Pathol.* **29(6):** 565–568.

68. Kanoe, H. *et al.* (1998) Amplification of the CDK4 gene in sarcomas: tumor specificity and relationship with the RB gene mutation. *Anticancer Res.* **18(4A):** 2317–2321.

69. Galanis, E. *et al.* (1998) Gene amplification as a prognostic factor in primary and secondary high-grade malignant gliomas. *Int. J. Oncol.* **13(4):** 717–724.

70. Jeffers, M. *et al.* (1997) Activating mutations for the met tyrosine kinase receptor in human cancer. *Proc. Natl. Acad. Sci. USA* **94(21):** 111445–111450.

71. Reifenberger, J. *et al.* (1998) Missense mutations in SMOH in sporadic basal

cell carcinomas of the skin and primitive neuroectodermal tumors of the central nervous system. *Cancer Res.* **58(9):** 1798–1803.

72. Sugimura, T. *et al.* (1990) Molecular biology of the hst-1 gene. *Ciba. Found. Symp.* **150:** 79–89.

73. Miller, W.H., Jr. *et al.* (1999) Reverse transcription polymerase chain reaction for the rearranged retinoic acid receptor alpha clarifies diagnosis and detects minimal residual disease in acute promyelocytic leukemia. *Proc. Natl. Acad. Sci. USA* **89(7):** 2694–2698.

74. Hunger, S.P. *et al.* (1998) E2A-PBX1 chimeric transcript status at end of consolidation is not predictive of treatment outcome in childhood acute lymphoblastic leukemias with a t(1;19)(q23;p13): a Pediatric Oncology Group study. *Blood* **91(3):** 1021–1028.

75. Munier, F.L. *et al.* (1997) Prognostic factors associated with loss of heterozygosity at the RB1 locus in retinoblastoma. *Ophthalmic Genet.* **18(1):** 7–12.

76. Berns, E.M. *et al.* (1998) Mutations in residues of TP53 that directly contact DNA predict poor outcome in human primary breast cancer. *Br. J. Cancer* **77(7):** 1130–1136.

77. Isaacs, W.B. (1995) Molecular genetics of prostate cancer. *Cancer Surv.* **25:** 357–379.

78. Frebourg, T. (1997) [Li-Fraumeni syndrome]. *Bull. Cancer* **84(7):** 735–740.

79. Schutte, M. *et al.* (1997) Abrogation of the Rb/p16 tumor-suppressive pathway in virtually all pancreatic carcinomas. *Cancer Res.* **57(15):** 3126–3130.

80. Tomlinson, I. *et al.* (1998) A comparison of the genetic pathways involved in the pathogenesis of three types of colorectal cancer. *J. Pathol.* **184(2):** 148–152.

81. Soravia, C., Bapat, B. and Cohen, Z. (1997) Familial adenomatous polyposis (FAP) and hereditary nonpolyposis colorectal cancer (HNPCC): a review of clinical, genetic and therapeutic aspects. *Schweiz. Med. Wochenschr.* **127(16):** 682–690.

82. Iacopetta, B.J. *et al.* (1998) Mutation of the transforming growth factor-beta type II receptor gene in right-sided colorectal cancer: relationship to clinicopathological features and genetic alterations. *J. Pathol.* **184(4):** 390–395.

83. Brassett, C. *et al.* (1996) Microsatellite instability in early onset and familial colorectal cancer. *J. Med. Genet.* **33(12):** 981–985.

84. Mautner, V.F. *et al.* (1998) [Neurofibromatosis versus schwannomatosis]. *Fortschr. Neurol. Psychiatr.* **66(6):** 271–277.

85. Richards, F.M. *et al.* (1998) Molecular genetic analysis of von Hippel-Lindau disease. *J. Intern. Med.* **243(6):** 527–533.

86. Claes, K. *et al.* (1999) Mutation analysis of the BRCA1 and BRCA2 genes results in the identification of novel and recurrent mutations in 6/16 flemish families with breast and/or ovarian cancer but not in 12 sporadic patients with early-onset disease. Mutations in brief no. 224. Online [In Process Citation]. *Hum. Mutat.* **13(3):** 256.

87. Unden, A.B. *et al.* (1997) Human patched (PTCH) mRNA is overexpressed consistently in tumor cells of both familial and sporadic basal cell carcinoma. *Cancer Res.* **57(12):** 2336–2340.

88. Saito, M. *et al.* (1998) Expression of DCC protein in colorectal tumors and its relationship to tumor progression and metastasis. *Oncology* **56(2):** 134–141.

89. Samowitz, W.S. *et al.* (1997) Beta-catenin mutations are more frequent in small colorectal adenomas than in larger adenomas and invasive carcinomas [In Process Citation]. *Cancer Res.* **59(7):** 1442–1444.

90. Saito, M. *et al.* (1999) Expression of DCC protein in colorectal tumors and its relationship to tumor progression and metastasis. *Oncology* **56(2):** 134–141.
91. Gryfe, R. *et al.* (1997) Molecular biology of colorectal cancer. *Curr. Probl. Cancer* **21(5):** 233–300.
92. Berx G. *et al.* (1996) E-cadherin is inactivated in a majority of invasive human lobular breast cancers by truncation mutations throughout its extracellular domain. *Oncogene* **13(9):** 1919–1925.
93. Samowitz, W.S. *et al.* (1999) Beta-catenin mutations are more frequent in small colorectal adenomas than in larger adenomas and invasive carcinomas [In Process Citation]. *Cancer Res.* **59(7):** 1442-1444.
94. Grulich, A.E. *et al.* (1999) Risk of cancer in people with AIDS [In Process Citation]. *AIDS* **13(7):** 839–843.
95. Chu, Z.L. *et al.* (1998) The tax oncoprotein of human T-cell leukemia virus type 1 associates with an persistently activates IkappaB kinases containing IKKalpha and IKKbeta. *J. Bio. Chem.* **273(26):** 15 891–15 894.
96. Butel, J.S. and Lednicky J.A. (1999) Cell and molecular biology of simian virus 40: implications for human infections and disease. *J. Natl. Cancer Inst.* **91(2):** 119–134.
97. De Luca, A. *et al.* (1997) The retinoblastoma gene family pRb/p105, p107, pRb2/p130 and simian virus-40 large T-antigen in human mesotheliomas [see comments]. *Nat. Med.* **3(8):** 913–916.
98. Kuo, K.W. *et al.* (1998) Variations in gene expression and genomic stability of human hepatoma cells integrated with hepatitis B virus DNA. *Biochem. Mol. Biol. Int.* **44(6):** 1133–1140.
99. Dawson, C.W. *et al.* (1998) Functional differences between BHRF1, the Epstein-Barr virus-encoded Bcl-2 homologue, and Bcl-2 in human epithelial cells. *J. Virol.* **72(11):** 9016–9024.
100. Farrell, P.J., Cludts, I. and Stuhler, A. (1997) Epstein-Barr virus genes and cancer cells. *Biomed. Pharmacother.* **51(6–7):** 258–267.
101. Teruya-Feldstein, J. *et al.* (1998) Expression of human herpesvirus-8 oncogene and cytokine homologues in an HIV-seronegative patient with multicentre Castleman's disease and primary effusion lymphoma. *Lab. Invest.* **78(12):** 1637–1642.
102. Duro, D. *et al.* (1999) Activation of cyclin A gene expression by the cyclin encoded by human herpesvirus-8. *J. Gen. Virol.* **80(pt. 3):** 549–555.
103. Kashanchi, F. *et al.* (1999) Human herpesvirus 6 (HHV-6) ORF-1 transactivating gene exhibits malignant transforming activity and its protein binds to p53. *Oncogene* **14(3):** 359–367.

Chapter 7

Nucleic acid analysis of historical samples

Nucleic acid analysis is not limited to the identification of disease mutations or infectious elements, it can also be used to answer questions of historical identity. In recent history, the claims of a Polish peasant woman to be Anastasia, the daughter of Nicholas II, Tsar of Russia, have been disproved by analysis of DNA from known descendants of the Romanoffs, while the remains of Tsar Nicholas II have been shown to be genuine. There are many other examples where DNA analysis has produced interesting information from ancient artefacts, and a few of these are discussed below.

7.1 The historical limits of DNA detection

The advent of PCR (see Section 2.3) brought with it claims for analysis of fossilized DNA from insects in amber about 100 million years old. Unfortunately these results have been shown to be artefactual, a finding that relegates Jurassic Park even more firmly into the realms of fantasy. Nonetheless it is possible to detect DNA in ancient tissue, as the work on frozen mammoths from Siberia has shown. These creatures were about 70 000 years old, but had the great advantage for those analyzing their DNA that they had been preserved frozen. Analysis of mitochondrial DNA showed similarity to that of Asian elephants. For bone, the most available tissue, the limits appear to be in tens of thousands of years, with studies on Neanderthal man those that have been successful for the oldest specimens.

7.1.1 The relationship between Neanderthal man and modern humans

Neanderthals lived in Europe and Western Asia for at least 100 000 years, and became extinct about 30 000 years ago. It has been a point of contention whether the Neanderthals were a separate species from *homo sapiens* or were in some way ancestral to them. By extremely careful analysis, in which every precaution was taken against contamination, it has been shown that mitochondrial DNA from Neanderthal bone differs markedly from human DNA from the same organelles. The changes suggest that the ancestors of humans and Neanderthals diverged about 600 000 years ago, and that Neanderthals did not contribute to the mitochondrial DNA of modern humans, i.e. they were separate rather than ancestral.

7.1.2 DNA analysis from mummified specimens

DNA has been examined from several mummies, including specimens from South America and Japan. Blood group genotype was established by PCR in nine mummified bodies in the Japanese study. In Colombian mummies, ranging from 150 to 1500 years in age, mitochondrial DNA was analyzed to show that the bodies were Amerindians with the same pattern of DNA as the existing populations.

7.2 Reasons for false results

The overwhelming reason for producing incorrect data in the analysis of ancient DNA is the presence of more recent material, either from the researcher or from the environment where the specimens were found. Very old bone specimens can often be contaminated by material from much more recent animals, even though there may be no visible sign of their presence. Studies on bones from the extinct South American ground sloth, which were about 10 000 to 20 000 years old, showed that only specimens found in a cold climatic region gave DNA that could be amplified, and even then this was a minority of samples. Extraction of DNA from 600 year old human remains from an archeological site in the USA showed that the relative amount of ancient to modern DNA in the sample was crucial. With less than 40 molecules to start with, several different sequences were produced. In contrast, a few thousand molecules as a starting point gave repeatable, unambiguous data. Oxidative decay of pyrimidine residues from six- to five-membered hydantoin rings is particularly damaging, as DNA polymerases are incapable of copying these altered residues.

7.3 Future possibilities

It may be that developments in DNA amplification technology will allow the copying of degraded bases in DNA, which is one of the main reasons that older and less well preserved specimens cannot be used at present. Even so, it seems that really ancient DNA, for instance from specimens trapped in amber, might never be available for analysis as the medium is excellent for preserving body structure, but not DNA. The work on Neanderthal man has shown that DNA analysis will be a useful tool in tracking the movement of human populations in the more recent past, though it sadly looks as though the recreation of dinosaurs will have to be left to Steven Spielberg. Perhaps, on the whole, that is not such a bad thing!

Further reading

Hoss, M., Dilling, A., Currant, A. and Paabo, S. (1996) Molecular phylogeny of the extinct ground sloth *Mylodon darwinii. Proc. Nat. Acad. Sci. USA* **93:** 181–185.

Ivanov, P.L., Wadhams, M.J., Roby, R.K., Holland, M.M., Weedon, V.W. and Parsons, T.J. (1996) Mitochondrial DNA sequence heteroplasmy in the Grand Duke of Russia Georgij Romanov establishes the authenticity of the remains of Tsar Nicholas II. *Nat. Genet.* **12:** 417–420.

Krings, M., Stone, A., Schmitz, R.W., Krainitzki, H., Stoneking, M. and Paabo, S. (1997) Neandertal DNA Sequences and the Origin of Modern Humans. *Cell* **90:** 19–30.

Lindahl, T. (1993) Recovery of antediluvian DNA. *Nature* **365:** 700.

Paabo, S. and Wilson, A.C. (1991) Miocene DNA sequences-a dream come true? *Curr. Biol.* **1:** 45–46.

Index